UNTAUGHT MATH

In Many High Schools

Math you may not have learned in High School.

Mathematics not learned even by the Curious High School Student including applications to cryptography

Math is the quest for absolute truth. If absolute truth exists at all, the study of mathematics comes closest to the quest.

SIDNEY KRIMSKY

B.S., M.S.M.E., MBA, P.E.
AIR COMMAND AND STAFF COLLEGE
AIR WAR COLLEGE

ISBN 978-1-950818-64-8 (paperback)

Copyright © 2020 by Sidney Krimsky

All rights reserved. No part of this publication may be reproduced, distributed, or transmitted in any form or by any means, including photocopying, recording, or other electronic or mechanical methods without the prior written permission of the publisher. For permission requests, solicit the publisher via the address below.

Rushmore Press LLC
1 888 733 9607
www.rushmorepress.com

Printed in the United States of America

Chapter i   Dedication

This book is dedicated to my two teachers at Abraham Lincoln High School in Brooklyn, New York; Mr. Ralph Ellis whose teaching method for plane and solid geometry was each day an exploration into the unknown and Mr. Benjamin Braverman who sat with me each day before school started and provided extra assignments for me to catch up for my inadequate background from Junior High School in algebra enabling me to excel in algebra, trigonometry and pre-calculus and inspired me to study mathematics in High School and then physics at the City College of New York. I regret that I never thanked them for their efforts. Both their names appear on the Abraham Lincoln Honor Roll.

I have to acknowledge the work done by the fine math teachers at the Huntington Learning Center in Lynbrook, New York who also impart the elegance and rigor of High School math to students who wish to excel in their high school studies. I learned from my students what is not taught in High School.

I have to acknowledge my wife Dorothy who manages our household affairs and social life making it possible for me to write this book. She is a loving wife, teacher, librarian, grandmother. great- grandmother, and counselor.

Chapter ii   What makes math so difficult?

The study of math is the only subject for which there are no opinions or opportunities to challenge the postulates, axioms, corollaries, deductions, and outcome of calculations. This is upsetting to many students who studied History, English, and Literature for which there can be many opinions and opportunities for self-expression.

Chapter iii   Purpose of this book

This book is not meant to be a text book or to replace any text book that demonstrates problem solving for algebra, geometry, trigonometry, or calculus. The purpose of this book is to enrich the minds of high school students studying math by presenting subjects that are not usually taught in high school math courses, but which are interesting and enjoyable for the curious minded high school students and their teachers.

Chapter iv   Credits:

Euclid
Greek Philosophers
Carl Friedrich Gauss
Leonardo Pisano Fibonacci
Leonhard Euler
George Kantor
Isaac Newton
Ibn al-Khawarizmi

Chapter v   Advantages to the brain by studying mathematics

In a study published in the *Proceedings of the National Academy of Sciences,* a pair of researchers at the ENSERM-CEA Cognitive Neuroimaging Unit in France reported that the brain areas involved in math are different from those engaged in equally complex nonmathematical thinking. The team used magnetic resonance imaging to scan the brains of 15 professional mathematicians and 15 non-mathematicians of the same academic standing,

While in the scanner the subjects listened to a series of 72 high-level mathematical statements, divided evenly among algebra, analysis, geometry, and topology, as well as 18 nonmathematical (mostly historical) statements. They had 4 seconds to reflect on each proposition and determine whether the proposition was true, false, or meaningless.

The researchers found that among the mathematicians only, listening to math related statements activated a network involving bilateral intraparietal, odorsal, prefrontal, and inferior temporal regions of the brain. This circuitry is usually not associated with areas involving language processing and semantics, which were activated in both mathematicians and non-mathematicians when they were presented with the nonmathematical statements, "On the contrary," says study co-author and graduate student Marie Amalric, "our results show that high level mathematical reflection recycles brain regions associated with an evolutionary ancient knowledge of number and space."

The following activities increase brain development by increasing the sizes of axons and dendrites between nerve cells: (Discover Magazine page 71, January/February 2019.)

- Learning languages at any age
- Music training especially playing or singing with others
- Gardening including learning, planning, artistry and physical work
- Crafts and sewing
- Learning dance and competitive work
- Games, crossword puzzles, playing chess
- Learning math and solving math problems
- Studying the Logic within the Talmud

Sid Krimsky tutored high school math students during high school. He graduated from CCNY with a degree in engineering physics. He tutored high school math part-time during the 1960s as a volunteer while working full time as an engineer. After retirement he joined the RE-SEED Program managed by the College of Education at Northeastern University in Boston in which about 90 STEM retirees volunteered one day per week to assist teachers and students within the Boston Public Schools. The STEM RE-SEED volunteers were motivated to share their valuable experience with students after having had a successful STEM career.

After three years with RE_SEED he moved to West Hempstead and tried unsuccessfully to bring the RE-SEED model to public and private schools in New York State. After two years he was hired by the Huntington Learning Center to teach high school math and science. During all that tutoring time he identified math subjects that are not taught to high school students which is the subject of this book.

Chapter vi Glossary of Terms used in this Book

ALICE

Sender of an encrypted email message to Bob

ASMAD

Adding, subtracting, multiplying and dividing. These are the fundamental arithmetic operations

BOB

Recipient of a an encrypted email message sent by Alice

CARDINALITY

Constitutes the number of elements in a set such as natural numbers, integers, irrational numbers, etc.

COMINT

Communications Intelligence includes interception and direction finding, radio finger printing, signal forensics and cryptanalysis.

COMSEC

Communications Security includes steganography (burying messages in noise), dummy messages, radio silence, and traffic messages, burst messages, and cryptography.

CRYPTOANALYSIS

is the art and science of code breaking to reveal the original plain text message. Synonyms are decoding and decrypting. Cryptographers and crypto-analysts are natural adversaries.

CRYPTOGRAPHIC KEY

is a a binary number that determines the functional output of a cryptographic algorithm. For encryption algorithms, a key specifies the transformation of plaintext into cipher text, and vice versa for decryption algorithms.

CRYPTOGRAPHY

Is the art and science, means, methods, and apparatus of converting or transforming plain text messages into encrypted messages and for reconverting the messages into their original plain text form by a simple reversal of the steps used in the transformation. Cryptography is about rendering plain information unintelligible and for restoring encrypted information to an intelligible form. A crypt is a place to hide or bury something or someone.

CRYPTOLOGY

is the art and science of the collection and/or exploitation of communications, products and the services to ensure integrity, authentication, confidentiality, and non-repudiation of electronic information. It embraces both cryptography and crypto-analysis. Cryptology includes Communication Security (COMSEC), Electronic Security (ELSEC), Communication Intelligence (COMINT), and Electronic Intelligence (ELINT).

DIFFERENTIAL GEOMETRY
Differential geometry is a mathematical discipline that uses the techniques of differential and integral calculus, as well as linear algebra, to study problems in non-linear curves and spaces

ELINT
Is electronic intelligence that includes electronic eavesdropping and countermeasures such as Jamming and false echoes and MASINT includes Measurement Intelligence, Data Collection, and Interpretation Signals Intelligence (SIGINT) = ELINT + COMINT

ELSEC
is electronic security and includes Counter-Counter measures to defeat jamming, EMSEC (to confuse an unauthorized listener), frequency shifting such as Time Division Multiple Access (TDMA).

EVE
Eve eavesdrops on messages sent by Alice to Bob or Bob to Alice

GEMATRIA
Is a system of assigning numbers to letter of the Hebrew Alphabet and deriving meaning from caparisons between paragraphs or events.

GEODESY
is the scientific discipline that deals with the measurement and representation of the Earth (or any planet), including its gravitational field, in a three-dimensional time-varying space.

HOMOPHONIC SHIFT
Each plane text letter in the alphabet is shifted by the same number to create a cypher text. This method is believed to have been used by Julius Caesar to communicate messages to his legions.

INFINITE SERIES
is a long series of fractions that may or may not converge

LOGARITHM BASE "e"
If $e^x = N$ then the logarithm of $N = x$. Logarithm of N is found in tables or hand calculators

LOGARITHM BASE "10"
If $10^x = N$ then the logarithm of $N = x$. Logarithm of N is found in tables or hand calculators

NEWTON
Is a measurement of force that equals 0.2248 lb.

NUMBER THEORY
Is a subset of mathematics that deals with the relationships between numbers and sometimes patterns in nature.

NUMERIC
is the study of the relationship between numbers that sometimes leads to surprising results. Numeric consists of numerology, number theory, and use of math that describes nature (physics).

NUMEROLOGY
is any belief in the divine or mystical relationship between a number and one or more coinciding events. It is also the study of the numerical value of the letters in words or names. Gematria is a subset of numerology.

PGP (PRETTY GOOD PRIVACY)
Cryptographic protection program for encrypted email transmissions

RE-SEED
Retirees Enhancing-Science Education through Experiments and Demonstrations

STEM
Science, Technology, Engineering and Mathematics

Untaught Math

(In many High Schools)

$$\sum_{I=1.0}^{N} i^3 = \left[\sum_{I=1.0}^{N} i \right]^2 = \left[\frac{N(N-1)}{2} \right]^2$$

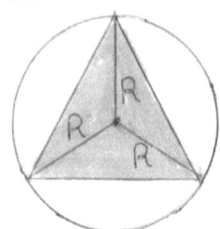  

Area Triangle 1   Area Triangle 2   Area Triangle 3

$$\frac{3/\sqrt{3}\,R^2}{4^1} \qquad \frac{3/\sqrt{3}\,R^2}{4^2} \qquad \frac{3/\sqrt{3}\,R^2}{4^3}$$

00011000 X 00000110 = 10010000

Sidney Krimsky

Contents

Chapter		Page Number
i	Dedication	iii
ii	What makes math so difficult?	iv
iii	Purpose of this book	iv
iv	Credits:	iv
v	Advantages to the brain by studying mathematics	v
vi	Glossary of Terms used in this Book	vii
1.0	Usefulness of numbers	1
2.0	Measurable Phenomena	1
3.0	Dimensional analysis	2
4.0	Roman Numbers	3
4.1	Roman Number Rules	4
4.2	Adding Roman Numbers	4
5.0	Natural Numbers used for counting	5
6.0	Integers and Numeric	6
6.1	Sum of Squares and Cubes of Sequential Integers.	6
6.2	Carl Friedrich Gauss	7
6.3	Sum of Sequential Natural Numbers	7
6.4	Horizontal Addition	8
6.5	Horizontal Subtraction	9
6.6	Horizontal Multiplication	9
6.7	Reversing Your Analog Wrist Watch:	10
6.8	Sum of Sequential Odd Numbers	10
6.9	ASMAD Fractions and Decimals	10
7.0	Rational Numbers	11
7.1	Convert a Fraction into a Decimal	12
7.2	Convert a Decimal into a Fraction	12
8.0	Irrational Numbers	13
8.1	Algebraic Real Numbers	14
8.2	Transcendental numbers	14
9.0	George Cantor 1845-1918	15
10.0	John Napier & Logarithms	15
10.1	Calculations using logarithms:	19
11.0	Ears and eyes are Logarithm detectors	20
11.1	Derivation of e (exponent)	23

11.2	Calculating value of "e" per Leonhard Euler using an infinite series 23
11.3	Compound Interest Annual Percentage Rate (APR) 24
11.4	Buying Interest Producing Bonds 24
11.5	Inflation (From Investopedia) 25
12.0	What is π? 28
12.1	Euler's Method Calculating π 28
12.2	Walls Method Calculating π 28
12.3	Rabbi Elijah of Vilna's method of computing π (1720-1797) 29
12.4	Archimedes Method to Calculate π. 30
12.5	Number of Integers in to test for repetition of integers 31
13.0	Imaginary Numbers 32
13.1	Complex Numbers contain real and imaginary numbers ... 33
13.2	Complex conjugate numbers 33
14.0	Solid geometry 33
14.1	Volume of Solids 34
14.2	Dihedral Angles 35
15.0	Vector addition 35
15.1	Multiplying Two Vectors (cross product) 37
15.2	Three Dimensional Vectors 37
15.3	Resultants of vectors 38
16.0	Complex number raised to a power 39
16.1	Graphical method to compute $(1 + i)^m$ 39
17.0	Tensors 40
17.1	Viscoelasticity Definition 42
18.0	Euler Series for calculating "e" and proving DeMoivre's Theorem. 42
19.0	Golden Ratio, Length/Width of Rectangle 43
20.0	Leonardo Pisano Fibonacci 45
20.1	Srinivasa Ramanujan 47
20.2	Golden Ratio for a pentagon 49
20.3	Fibonacci Patterns 49
20.4	Quadratic Equation Solution & Golden Ratio 51
20.5	Dr. Robert Weller's Axiom 52
20.6	Zeno's paradox or you can't get there from here 53

20.7	Infinite Series Convergence	54
20.8	Casting out Nines – First Method	54
20.9	Casting out nines-Second Method	55
30.0	Two Digit Reversal Math	55
30.1	Three Digit Reversal Math	55
30.2	Five regular polyhedrons	56
30.3	Regular Polygons (side or radius = 1.0)	58
40.0	Binary Numbers Gottfried Wilhelm Leibnitz	58
41.0	Prime Numbers and Cryptography	59
41.1	Fermat's Test for Prime Numbers	63
41.2	Algorithm for generating prime numbers.	63
42.0	Perfect Numbers	66
43.0	Smith Numbers	68
44.0	Magic Numbers	69
44.1	Magic Calculations	70
45.0	Solving Linear Simultaneous Equations	71
45.1	Solving Two-Dimensional Linear Simultaneous Equations by Substitution	71
45.2	The second method involves subtraction	72
45.3	The third method involves linear algebra and matrices	73
45.4	Determinant = 0, For Parallel Lines	74
45.5	Solution if slopes are of opposite signs.	74
45.6	Graphical Solution	75
45.7	Solving 3 dimensional linear equations	76
45.8	Example of solving an electrical circuit with 3 unknowns	79
46.0	Dropping a stone through the earth	85
47.0	Computer Accuracy Check:	91
48.0	Quiz	91
49.0	Cryptography Introduction	92
49.1	Time needed to open a 6 number combination lock	94
49.2	Example:	94
49.3	Frequency analysis for English Language:	96
49.4	Leslie Groves Cypher	98
49.5	Polyphonic Substitution	99
49.6	Random Letter Substitution	100
49.7	Thomas Jefferson's Cypher Wheel	101

50.0	Combination and Permutations Reminder:	104
50.1	Word Counts	106
50.2	Letter Counts	107
50.3	Random Number Homophonic Substitution	109
60.0	Binary Numbers	110
60.1	ASMAD for binary numbers	112
60.2	ASCII	114
60.3	Reversible Algorithms	116
60.4	Reversible and Irreversible algorithms	117
60.5	Swapping BITS	119
60.6	IBM Encryption Technology	120
60.7	Diffie-Hellman Amazing Discovery	122
61.0	Modulo Math and Irreversible Algorithms	124
61.1	RSA Algorithm Public & private key cryptography	127
61.2	Calculations using Modulo Math-Table I	131
70.0	Prime Numbers Needed for Cryptography	132
70.1	Privacy and National Security	133
70.2	Origins of PGP (Pretty Good Privacy)	133
70.3	Politics of Protection	135
71.0	Quantum Cryptography (QC)	138
71.2	MagiQ Technologies	144
71.3	Recent Developments in Quantum Cryptography	145
71.4	Quantum Entanglement	148
72.0	Geometry Rhombus Smallest Perimeter for a parallelogram	149
72.1	Smallest perimeter for a Rhombus	150
72.2	Ambiguous Case for Proving Triangles Congruent	150
72.3	Laws of Sine's and Cosines for any triangle	152
72.4	Finding the 3rd Side of a Triangle Using the (Law of Cosines)	152
72.5	Perfect Right Triangles (called triplets)	153
73.0	Global Positioning System (GPS)	154
73.1	Time Dilation Due to Speed	157
73.2	Time Dilation due to Gravity	158
73.3	Effect of Speed and Gravity on GPS Accuracy	159
74.0	Newton's Laws	160

74.1	Newton's Law of Gravity	163
74.2	Centrifugal and centripetal forces.	164
75.0	Growth Models	166
76.0	Probability	167
76.1	Birthday Problem (From Wikipedia)	167
76.2	Probability of Achieving Various Hands at Poker	168
76.3	Statistics	169
77.0	Set Theory	174
78.0	Equal Product and Sum of Two Integers	175
78.1	"Eight This Cool" (Jewish Home Magazine) Page 82, 6 Dec 2018	176
78.2	Cos (Cos (Cos (Cos Θ))) etc. = 0.7389	176
78.3	Asymptotes	178
78.4	Asymptotes for a Hyperbola	186
79.0	Finding the Areas of SAS, ASA, and SSS Triangles.	187
80.0	Completing the Square for Solving Quadric Equations	188
81.0	Length of a Chord	189
81.1	Area of a Segment of a Circle	190
81.2	Centroid of a Triangle	191
81.3	Length of Medians of a Triangle	193
81.4	Inequalities and Graphical Display	194
81.5	Resultant of Adding Two Vectors Using Law of Cosines	197
81.6	Resultant of Subtracting Two Vectors Using Law of Cosines	198
81.7	Area and Perimeter of an Ellipse.	198
82.0	Collatz Problem	199
83.0	Quadratic Equations	199
83.1	Parabola	200
83.2	Derivatives for Parabola	204
83.3	Factoring $1000C^3 - 27$	205
83.4	Adding integers in a sequence	207
84.0	References	208

Chapter 1.0 Usefulness of numbers

Numbers are used to measure and compare physical phenomena by first establishing standards. There only five basic standards for measurement

1. Distance (including area, volume) feet, inches, and angular measurement
2. Time (including seconds, hours, years, etc.)
3. Force (including weight) in lbs., or newtons.
4. Electric charge in coulombs:
   - coulombs/sec = amps,
   - volts = joules/coulomb or newton-meters/coulomb.
   - power = volts × amps = watts,
   - watt-seconds = energy, (converts to foot-lbs. or newton meters)
5. Counting (cardinal numbers), placement (ordinal numbers such as first, second, third…)& labeling
6. Images are pixelated and assigned a number from 0 to 256 (2^8) depending on the optical density of the image using a densitometer.

Chapter 2.0 Measurable Phenomena

Measurable phenomena employ the five uses of numbers to describe the world

Examples:

1. Temperature may be measured by a linear expansion of a liquid in a narrow tube or a bi-metallic strip or of a change of properties in a microchip

2. Time is used for all rate changing phenomena such as:
   - velocity, = distance/time
   - acceleration = velocity /time
   - jerk = acceleration/time

3. Magnetism is a force from crystal alignment of certain metals or a moving electric charge or from a metal bar wrapped with an electric wire

4. Gravity is a force between masses that attract each other according to Isaac Newton

5. Area and volume are distances squared and cubed

6. Voltage is joules/coulomb

7. Mass is not measurable; a mass meter does not exist. Mass is inferred from Newton's second Law; $F = m\ a$,; where F = force, a = acceleration. Therefore; mass = lbs./ft/sec^2 = 1 slug. Mass is an invented term to satisfy Newton's Laws of Motion and Gravity.

"To measure is to know" Lord Kelvin.

Chapter 3.0   Dimensional analysis

Dimensional analysis is necessary but not sufficient to validate equations describing physical phenomena. Dimensions (such as force, time, distance, charge, etc.) must be equal on both sides of the equation. Example (1)

A person paints 5/6 of a fence in 1 hour. How long to paint the entire fence?

1 Fence = [5/6 Fence/hour] × time

Divide both sides by 5/6 fence/hour

1 fence/(5/6 fence/hour) = time

After canceling the word fence, and dividing we get: 6/5 hour = time = 1.2 hours

Dimensions such as length, time, distance, & force etc. are treated as items that can be canceled If they appear on both sides of an equation. The dimensions are treated as variables.

Example (2)

What is the pressure on a diver under 10 feet of water?

Water weighs 62.4 lbs./foot^3 pressure = density (lbs/ft^3 × height (ft)

The pressure of water on the diver = 62.4 lbs./foot^3 × 10 feet = 624 lbs./foot^2 / 144 inches^2/foot^2. The terms foot^2 are cancelled because it appears in the numerator and denominator.

= 4.33 lbs./inches^2 or 4.33 psi.

Chapter 4.0 Roman Numbers

I = 1   C = 100
V = 5   D = 500
X = 10   M = 1000
L = 50

Chapter 4.1 Roman Number Rules

Letters before numbers imply subtraction

IV = 5 – 1 = 4

IX = 10 – 1 = 9

CD = 500 – 100 = 400

Letters after numbers imply addition

VI = 5 + 1 = 6; VIII = 8

XI = 10 + 1 = 11; XIII = 13

DC = 500 + 100 = 600; DCCC = 800

Chapter 4.2 Adding Roman Numbers

LVI + XCII + XVI = 56 + 92 + 16 = 164 = CLXIV

Sort by letters and replace XC by LXXXX (100 – 10) = 50 + 10 + 10 + 10 + 10)

L		V	I
L	XXXX	II	
	X	V	I

LL XXXXX   VV   IIII = LL XXXXX VV IIII

CLXIIII = 164

ASMAD is cumbersome compared to our decimal system. (ASMAD = addition, subtraction, multiplication and division)

Chapter 5.0 Natural Numbers used for counting

1, 2, 3, 4, 5, 6, 7, etc……. are sometimes called natural numbers. These are whole numbers or cardinal numbers.

Zero is excluded

ASMAD is limited because zero and negative numbers are excluded.

Division is a series of successive subtractions:

Example: 15/4   15 - 4 = 11 first subtraction

11 - 4 = 7 second subtraction

7 - 4 = 3, third subtraction, 3 < 4

Therefore 15/4 = 3 + R3; "R" is remainder

Results with "R" are not additive meaning that

13/4 = 3 + R1 = 3.25

15/4 = 3 + R3 = 3.75

15/4 + 13/4 = 7.00

3 + R1 + 3 + R3 does not = 6 + R4; 6 + R4 = 10

The set of natural numbers are infinite.

Chapter 6.0 Integers and Numeric

Numbers ....-4, -3, -2, -1, 0, 1, 2, 3, 4, 5, ...etc. are called integers. Zero and negative numbers are included.

ASMAD division is cumbersome just as with natural numbers.

Therefore 15/4 = 3 + R3; R is remainder

Fractions are excluded.

Zero was discovered by Aryabhatl in India around year 900 BCE.

Mohammed ibn al-Khawarizmi is credited with inventing algebra.

The word "algorithm" is a corruption of his name.

According to Prof George Kantor, the cardinality (total sum) of the set of integers equals the cardinality of the set of natural numbers.

Chapter 6.1 Sum of Squares and Cubes of Sequential Integers.

The sum of a sequence of integers squared equals the sum of the cubes of each integer. Example:

$(1 + 2 + 3 + 4 + 5 + 6 + 7 + 8)^2 = 1296$

$1^3 + 2^3 + 3^3 + 4^3 + 5^3 + 6^3 + 7^3 + 8^3 = 1296$

This relationship extends to all integers.

Chapter 6.2 Carl Friedrich Gauss

Carl Friedrich Gauss was one of the greatest mathematician who ever Lived. Asked at 10 years old to calculate sum of numbers from 1 to 100 he quickly said 5050. The formula he used was as follows:

If n = 100, then $\{[n](n+1)/2\} = 5050$

Similarly, the sum of sequential numbers from 1 to 10 = 55. How did he do it?

Use method of Karl Friedrich Gauss at age of 10.

Chapter 6.3 Sum of Sequential Natural Numbers

Add two rows of numbers

1+ 2 + 3 + 4 +   5+ 6+ 7 + 8+ 9 + 10

10+ 9 + 8 + 7 +   6+ 5+ 4 + 3+ 2 + 1

11+11+11+11+   11+ 11+ 11+ 11+ 11+ 11

Therefore: Answer = 10 numbers × sum/2

Since there are two rows: 10 × 11/2 = 55

Gauss formula: Sum = N(N+1)/2, N is the last integer.

Sum of first 100 numbers = 100X101/2 = 5050

Gauss' contribution to mathematics includes

- Number theory
- Differential geometry
- Statistics
- Probability –bell shaped or Gaussian curve
- Geodesy - measurements for non-Euclidian geometry
- Electromagnetism-magnetic and electric field theory
- Astronomy – motion of celestial objects
- Linear algebra ASMAD
- Regression Analysis (used to establish coorelations)
- Graphic representation of complex numbers

Chapter 6.4   Horizontal Addition

Suppose we have to add a column of integers:

Add the numbers vertically and horizontally

3578	$3 + 5 + 7 + 8 = 23$	$2 + 3 = 5$
5419	$5 + 4 + 1 + 9 = 19$	$1 + 9 = 10$
<u>5283</u>	<u>$5 + 2 + 8 + 3 = 18$</u>	<u>$1 + 8 = 9$</u>
14280	$1 + 4 + 2 + 8 + 0 = 15$	$5 + 10 + 9 = 24$
	$1 + 5 = 6$	$2 + 4 = 6$
	$6 = 6$	

The sum of $14280 - 6$ is always divisible by 9.

$(14280-6)/9 = 1586$. This idea is always true.

Chapter 6.5 Horizontal Subtraction

67519	6 + 7 + 5 + 1 + 9 = 28	2 + 8 = 10
(54321)	5 + 4 + 3 + 2 + 1 = 15	1 + 5 = (6)
13198	1 + 3 + 1 + 9 + 8 = 22	10 - 6 = 4
	2 + 2 = 4	4 = 4

The answer of 13198 - 4 = 13194. 13194/9 = 1466

The answer minus the sum of the integers in the answer is always divisible by 9.

Chapter 6.6 Horizontal Multiplication

357	3 + 5 + 7 = 15	1 + 5 = 6
× 692	6 + 9 + 2 = 17	1 + 7 = 8
247044	6 × 8 = 48	4 + 8 = 12; 1 + 2 = 3
2 + 4 + 7 + 0 + 4 + 4 = 21	2 + 1 = 3	

The answer 247044–3 is divisible by 9.

If horizontal multiplication does not produce the same number (3) then the answer is incorrect. Here is another example:

5316	5 + 3 + 1 + 6 = 15	1 + 5 = 6		
× 34	3 + 4 = 7		7 × 6 = 42	4 + 2 = 6
180744	1 + 8 + 0 + 7 + 4 + 4 = 24			2 + 4 = 6

Chapter 6.7 Reversing Your Analog Wrist Watch:

If you turn your wristwatch upside down, the handles always show 6 ½ hours ahead. Add 30 minutes, and that is the time in Israel. Israel is 7 hours ahead of the Eastern Time Zone.

Chapter 6.8 Sum of Sequential Odd Numbers

$1 = 1 = 1^2$

$1 + 3 = 4 = 2^2$

$1 + 3 + 5 = 9 = 3^2$

$1 + 3 + 5 + 7 = 16 = 4^2$

$1 + 3 + 5 + 7 + 9 = 25 = 5^2$

Sum of the first n odd numbers = n^2

Chapter 6.9 ASMAD Fractions and Decimals

ASMAD now possible for fractions, assume that a, b, c, d, e, & f are real numbers

Adding two Fractions:

a/b + c/d = (ad + bc)/bd

Adding three Fractions:

$a/b + c/d + e/f = (fad + fbc + ebd)/bdf$

Subtracting two Fractions:

$a/b - c/d = (ad - bc)/bd$

Multiplying two Fractions:

$a/b \times c/d = ac/bd$

Dividing two Fractions:

$(a/b)/(c/d) = a/b \times d/c$

Chapter 7.0   Rational Numbers

Rational numbers include integers with fractions can be expressed as the ratio of two integers. Numerator and denominators are integers (excluding zero in the denominator) includes repeating decimals such as:

$0.33333\ldots\ldots = 1/3$;   or 0.142857142857 etc. … $= 1/7$

$-1/2, -5/16, 3/17, 0.3125 = 5/16$ are examples of rational numbers

Fractions that may be expressed as the ratio of two integers are rational numbers

The cardinality of rational numbers equals the cardinality of integers per George Cantor.

Chapter 7.1 Convert a Fraction into a Decimal

Fractions can always be converted into decimals by dividing the denominator into the numerator.

Example 5/16 = 0.3125. Converting 0.3125 into a fraction is not easy. Decimals can be converted into fractions by many hand-held calculators.

Here is one way to do this: Example: 0.3125 =

3/10 + 1/100 + 2/1000 + 5/10,000 = 3/10 + 1/100 + 1/500 + 1/2000

2,000 is the least common denominator

Multiply denominator and numerator by 2,000

2000 [3/10 + 1/100 + 1/500 + 1/2000] = (600 +20+ 4 +1)/2000 = 625/2000 = 0.3125

Chapter 7.2 Convert a Decimal into a Fraction

Convert 0.3125 into a fraction:

0.3125 = X/Y

Select Y's until (0.3125) × (Y) = an integer.

Y > X

Example:

Y = 12 ,then 0.3125 × 12 = 3.750   no integer

Y = 13, then 0.3125 × 13 = 4.06   no integer

y = 14, then 0.3125 × 14 = 4.375   no integer

y = 15, then 0.3125 × 15 = 4.687   no integer

Y = 16, then 0.3125 × 16 = 5.000   an integer

STOP

Therefore 0.3125 = 5/16

Chapter 8.0   Irrational Numbers

Square roots, cube roots etc. Mixed with rational numbers are called irrational numbers

$\sqrt{2}$ = 1.414213562……etc.…

Integers describing irrational numbers do not repeat. Natural numbers, Integers, Rational, and Irrational Numbers are called complex numbers. The set of real numbers consists of natural numbers, integers, rational numbers and irrational numbers. Later we shall add algebraic real numbers and transcendental numbers.

Chapter 8.1 Algebraic Real Numbers

Algebraic real or complex numbers represent the solution to a polynomial equation with integer coefficients and exponents. For example: X = an algebraic real or complex number because X is a solution to the polynomial equation

$$AX^5 + BX^3 + CX + D = 0.$$

The set of algebraic numbers exceed the set of rational; numbers according to George Cantor.

Chapter 8.2 Transcendental numbers

Transcendental numbers are never the solutions to polynomial equations with integer coefficients.

Can π be a solution to a quadratic equation? Assume that π is a solution

$$\pi = \frac{-b \pm \sqrt{b^2 - 4ac}}{2a}$$

Let a = 1 & b = 5 (as an example) Then

$$2\pi + 5 = \sqrt{25 - 4c}$$

$$4\pi^2 + 20\pi + 25 = 25 - 4C$$

$$C = -\pi^2 - 5\pi$$

C is not an integer, it is a transcendental number

Therefore π is not a solution to a quadratic equation with integer coefficients. Integers, rational numbers, irrational numbers, logarithms, algebraic numbers, transcendental numbers, Sine's, Cosines, and tangents of angles are all called real numbers.

Chapter 9.0   George Cantor 1845-1918

George Cantor proved that the set of transcendental numbers exceeds the set of algebraic numbers.

Georg Cantor – his view: The Infinite Book: "A short Guide to the Boundless, Timeless, and Endless…" started to tell his friends that he had not been the inventor of the ideas about infinity that he had published. He was merely a mouthpiece, inspired by God to communicate parts of the mind of God to everyone else.

Chapter 10.0 John Napier & Logarithms

John Napier 1550 -1617, invented logarithms in 1594 to assist astronomers in making long calculations. Napier's bones became precursor to the slide rule. Slide rules and logarithms have been replaced by microchips in calculators.

Definitions:

There are two bases for logarithms:

Base "e" and base "10"; "e" means exponent. Base 10 is the decimal system

Logarithm of 1,000 (base 10) means that if $10^X = 1,000$ then $X = 3.0$

Logarithm of 1,000 (base e) means that if $2.7184^X = 1,000$ then $X = 6.90775$....

Logarithms to base "e" are called natural logarithms [e = 2.71828182845904...etc.].

Logarithms to base 10 are called common logarithms

Shown on page 19 is a circular and longitudinal slide rule. The lengths of the scales are proportional to the logarithms of numbers. Multiplying two numbers is achieved by adding the numbers on the scale and reading the anti-log. The slide rule also has sine's, cosines and tangents of angles as well as their hyperbolic functions.

Logarithms are used for multiplication, division, exponentiation, and raising numbers to positive and negative exponents. Table of logarithms of numbers from 1 to 1,000 were available in high school text books. However, the operations are replaced by algorithms burnt onto microchips available in hand-held calculators. Slide rules and logarithm table are now museum relics.

The first logarithm tables was compiled by Henry Briggs and published in Prague in 1629.

The logarithm of any number consists of a characteristic and a mantissa.

Example: What is the natural logarithm of 4870

The number 4800 may be expressed as 48.70×10^2

The natural logarithm = log(n) of 4870 = log 48.70 + Log(n) 100
= 3.8856 + 4.605 = 8.4908

This means that $e^{8.4908} = 4870$

Using common logarithms (base 10)

Ln (10) = 2.302585093

Common Logarithm of 4870 = Common Log of (48.7 X 100)

= Common Log of 48.7 + Common Log of 100

= 1.687529 + 2.000000 = 3.687529. Therefore $10^{3.6875} = 4870$

Note that 8.4908/2.302585093 = 3.6875, same as above.

A Table of Natural Logarithms followed by a linear and circular slide rule are shown on the next two pages.

UNTAUGHT MATH 17

Natural Logarithm Table

N	0	1	2	3	4	5	6	7	8	9
1.0	0.0000	0.0100	0.0198	0.0296	0.0392	0.0488	0.0583	0.0677	0.0770	0.0862
1.1	0.0953	0.1044	0.1133	0.1222	0.1310	0.1398	0.1484	0.1570	0.1655	0.1740
1.2	0.1823	0.1906	0.1989	0.2070	0.2151	0.2231	0.2311	0.2390	0.2469	0.2546
1.3	0.2624	0.2700	0.2776	0.2852	0.2927	0.3001	0.3075	0.3148	0.3221	0.3293
1.4	0.3365	0.3436	0.3507	0.3577	0.3646	0.3716	0.3784	0.3853	0.3920	0.3988
1.5	0.4055	0.4121	0.4187	0.4253	0.4318	0.4383	0.4447	0.4511	0.4574	0.4637
1.6	0.4700	0.4762	0.4824	0.4886	0.4947	0.5008	0.5068	0.5128	0.5188	0.5247
1.7	0.5306	0.5365	0.5423	0.5481	0.5539	0.5596	0.5653	0.5710	0.5766	0.5822
1.8	0.5878	0.5933	0.5988	0.6043	0.6098	0.6152	0.6206	0.6259	0.6313	0.6366
1.9	0.6419	0.6471	0.6523	0.6575	0.6627	0.6678	0.6729	0.6780	0.6831	0.6881
2.0	0.6931	0.6981	0.7031	0.7080	0.7129	0.7178	0.7227	0.7275	0.7324	0.7372
2.1	0.7419	0.7467	0.7514	0.7561	0.7608	0.7655	0.7701	0.7747	0.7793	0.7839
2.2	0.7885	0.7930	0.7975	0.8020	0.8065	0.8109	0.8154	0.8198	0.8242	0.8286
2.3	0.8329	0.8372	0.8416	0.8459	0.8502	0.8544	0.8587	0.8629	0.8671	0.8713
2.4	0.8755	0.8796	0.8838	0.8879	0.8920	0.8961	0.9002	0.9042	0.9083	0.9123
2.5	0.9163	0.9203	0.9243	0.9282	0.9322	0.9361	0.9400	0.9439	0.9478	0.9517
2.6	0.9555	0.9594	0.9632	0.9670	0.9708	0.9746	0.9783	0.9821	0.9858	0.9895
2.7	0.9933	0.9969	1.0006	1.0043	1.0080	1.0116	1.0152	1.0188	1.0225	1.0260
2.8	1.0296	1.0332	1.0367	1.0403	1.0438	1.0473	1.0508	1.0543	1.0578	1.0613
2.9	1.0647	1.0682	1.0716	1.0750	1.0784	1.0818	1.0852	1.0886	1.0919	1.0953
3.0	1.0986	1.1019	1.1053	1.1086	1.1119	1.1151	1.1184	1.1217	1.1249	1.1282
3.1	1.1314	1.1346	1.1378	1.1410	1.1442	1.1474	1.1506	1.1537	1.1569	1.1600
3.2	1.1632	1.1663	1.1694	1.1725	1.1756	1.1787	1.1817	1.1848	1.1878	1.1909
3.3	1.1939	1.1969	1.2000	1.2030	1.2060	1.2090	1.2119	1.2149	1.2179	1.2208
3.4	1.2238	1.2267	1.2296	1.2326	1.2355	1.2384	1.2413	1.2442	1.2470	1.2499
3.5	1.2528	1.2556	1.2585	1.2613	1.2641	1.2669	1.2698	1.2726	1.2754	1.2782
3.6	1.2809	1.2837	1.2865	1.2892	1.2920	1.2947	1.2975	1.3002	1.3029	1.3056
3.7	1.3083	1.3110	1.3137	1.3164	1.3191	1.3218	1.3244	1.3271	1.3297	1.3324
3.8	1.3350	1.3376	1.3403	1.3429	1.3455	1.3481	1.3507	1.3533	1.3558	1.3584
3.9	1.3610	1.3635	1.3661	1.3686	1.3712	1.3737	1.3762	1.3788	1.3813	1.3838
4.0	1.3863	1.3888	1.3913	1.3938	1.3962	1.3987	1.4012	1.4036	1.4061	1.4085
4.1	1.4110	1.4134	1.4159	1.4183	1.4207	1.4231	1.4255	1.4279	1.4303	1.4327
4.2	1.4351	1.4375	1.4398	1.4422	1.4446	1.4469	1.4493	1.4516	1.4540	1.4563
4.3	1.4586	1.4609	1.4633	1.4656	1.4679	1.4702	1.4725	1.4748	1.4770	1.4793
4.4	1.4816	1.4839	1.4861	1.4884	1.4907	1.4929	1.4951	1.4974	1.4996	1.5019
4.5	1.5041	1.5063	1.5085	1.5107	1.5129	1.5151	1.5173	1.5195	1.5217	1.5239
4.6	1.5261	1.5282	1.5304	1.5326	1.5347	1.5369	1.5390	1.5412	1.5433	1.5454
4.7	1.5476	1.5497	1.5518	1.5539	1.5560	1.5581	1.5602	1.5623	1.5644	1.5665
4.8	1.5686	1.5707	1.5728	1.5748	1.5769	1.5790	1.5810	1.5831	1.5851	1.5872
4.9	1.5892	1.5913	1.5933	1.5953	1.5974	1.5994	1.6014	1.6034	1.6054	1.6074
5.0	1.6094	1.6114	1.6134	1.6154	1.6174	1.6194	1.6214	1.6233	1.6253	1.6273
5.1	1.6292	1.6312	1.6332	1.6351	1.6371	1.6390	1.6409	1.6429	1.6448	1.6467
5.2	1.6487	1.6506	1.6525	1.6544	1.6563	1.6582	1.6601	1.6620	1.6639	1.6658
5.3	1.6677	1.6696	1.6715	1.6734	1.6752	1.6771	1.6790	1.6808	1.6827	1.6845
5.4	1.6864	1.6882	1.6901	1.6919	1.6938	1.6956	1.6974	1.6993	1.7011	1.7029

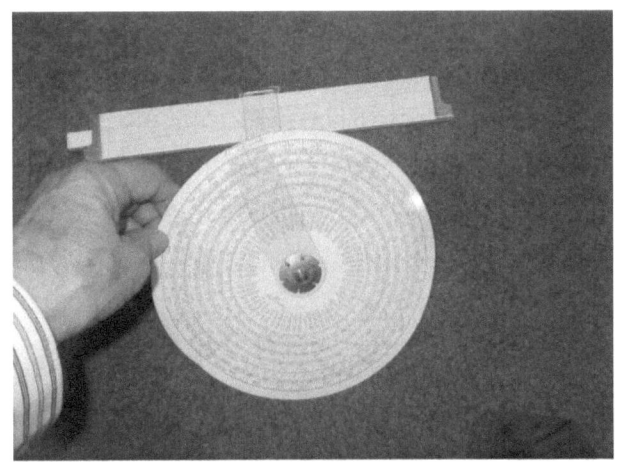

Chapter 10.1 Calculations using logarithms:

Suppose we have to calculate A × B/C = D where A, B, C & D are large numbers.

So we can say: Log A + Log B – Log C = Log D

Assume: A = 2345450, B = 9865476 and C = 5432089

From Common Log Tables we can find the common logarithms of the three numbers.

Log A = 6.370226,   Log B = 6.994118&   Log C = 6.734966

Log A + Log B - Log C = 6.629378, $10^{6.629379}$ = 4259690.055

Multiplying: 2345450 × 9865476 /5432089 = 4259683.647

The difference between the two answers is:
6.408 out of 4259683 = 1.5×10^{-4} %

This is a tedious calculation using logarithms. The length of a slide rule is proportional to the logarithms of numbers. Slide rules can only obtain an answer to three significant integers. The answer on my slide rule is 4.22. Engineers built bridges based on slide rule calculations. The longitudinal slide rule cost $24 in 1956. A hand-held calculator with memory and more functions than a slide rule, today costs around $10. Hand calculators became available around 1970.

Chapter 11.0 Ears and eyes are Logarithm detectors

Minimum detectable sound level: 1.0×10^{-12} watts/cm^2

Minimal detectable vision level: 4.8×10^{-12} watts/cm^2

The eyes are protected by the iris that closes when light entering the eyes is excessive. However, the reaction time is slow and the eye may be damaged by exposure to excessive light. The ears also have such protection.

Sound decibels are defined as:

$$\frac{10 \times \log (\text{Base 10}) \text{ of Power delivered in watts/ cm}^2}{10^{-12} \text{ watts/ cm}^2}$$

Maximum allowed decibels before hearing loss = 90 decibels (continuous for 8 hours)

Light decibels are defined as:

Maximum light received before eye damage: 30 milliwatts/cm2

$$\frac{10 \times \log \text{(Base 10) } 0.03 \text{ watts/cm}^2 \text{ (from sun)}}{4.8 \times 10^{-12} \text{ watts/cm}^2}$$

Maximum allowable decibels before eye damage:

$= 10 \times \log \text{(Base 10)}(3.0/4.8) \times 10^{10} = 20.4$ decibels (at a frequency of 442 nanometers)

Protection for eyes is the iris that acts as a shutter that closes in response to bright light to avoid damaging the retina. Protection for the ears is accomplished by muscles that tighten the eardrum. However reaction time for both the eye and eardrum are slow and a bright light or loud noise can still damage the eye and ear. The following diagrams show how the ear is protected from loud noise.

Sunglasses (especially polarized sun glasses) provide protection for eyes and earplugs provide protection for ears.

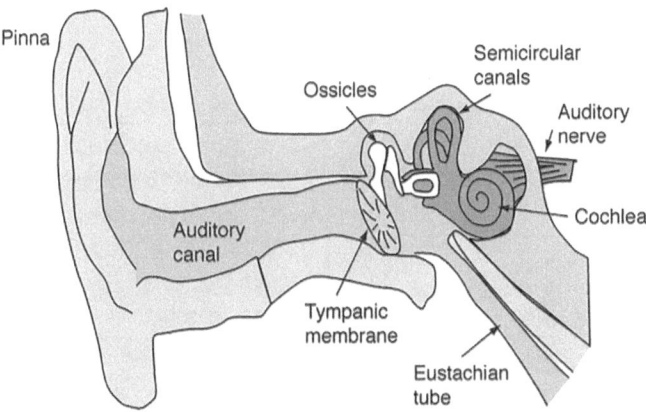

Loud Sound Response

In response to loud sounds, the tensor tympani muscle tightens the eardrum and through the tendon between the hammer and anvil and shifts the stirrup backward from the oval window of the inner ear. This shifting of the ossicles reduces the transmitted force to the inner ear, protecting it. However, it is a relatively slow action and cannot protect the ear from sudden loud sounds like a gunshot. The process cannot protect against sudden loud noises.

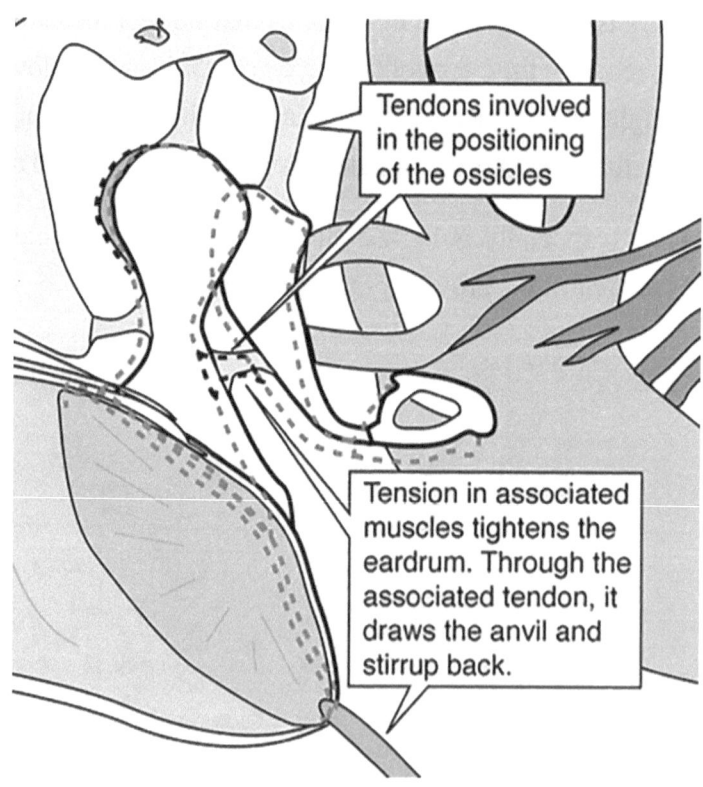

Chapter 11.1 Derivation of e (exponent)

e = limit of $(1 + 1/n)^n$ as n approaches infinity ;

For example Let n = 1000, then e = 2.716923932...

Let n = 10,000, then e = 2.718269596...

Actual e = 2.71828182845904...Transcendental integers do not repeat; e is transcendental.

One dollar earning 1.0% with infinite compounding, becomes $2.72

Chapter 11.2 Calculating value of "e" per Leonhard Euler using an infinite series

e = 1+ 1/1 + 1/(1X2) + 1/(1X2X3) + 1/(1X2X3X4) + 1/(1X2X3X4X5) ... etc.......

e = 2 + 1/2 + 1/6 + 1/24 + 1/120 +... etc...

e = 2.716666......

Thanks to Leonhard Euler. e means exponent and appears in math that involves growth.

Chapter 11.3 Compound Interest Annual Percentage Rate (APR)

One dollar invested at 4% would reach $1.49 after 10 years if compounded 4 times each year.

The equation is: Principal (1 + Interest rate/4 times per year) raised to the 40th power because 10 years × 4 times per year = 40

$$\$1.49 = \{1 + 0.04/4\}^{40}$$

4% interest compounded 4 times in one year is called the Annual Percentage Rate (APR) for 4%.

$$\{(1 + 0.04/4)^4 - 1\} \times (100) = APR = 4.06\%$$

Chapter 11.4 Buying Interest Producing Bonds

In general, there are two kinds of bonds.

First kind: Interest is paid to purchaser four times per year or once a year.

Second kind: Interest is allowed to accumulate until bond is sold or surrendered and the purchaser receives payment in excess of the purchased price because the purchase price includes interest earned. The difference between sales price and purchased price is called a "capital gain" and the capital gain is taxable on federal and state tax returns in the year in which the purchaser receives the interest payment.

Example 1:

Purchaser buys a bond for $1,000 and receives 4% interest every 3 months of the year. The interest received every 3 months = $1,000 × 0.04/4 = $10.00 Total interest = $40 per year. The capital gain each year of $40 is subject to state and federal taxes.

Example 2:

Purchaser buys a bond for $1,000 and allows interest to be re-invested at 4% per Year for 5 years at the end of which time the purchaser sells the bond back to the bank for the purchase price plus accumulate interest. Sales Price 1 = $1,000 $\{1 + 0.04/4\}^{40}$ = $1,488.86 (4 The exponent of 40 = 4 times each year × 10 years. Purchase Price = $1,000 × 1.4886 = $1,488.86. The capital gain in this case is $488.86 and is subject to state and federal taxes.

Chapter 11.5 Inflation (From Investopedia)

The measurement of inflation is a difficult problem for government statisticians. To do this, a number of goods that are representative of the economy are put together into what is referred to as a "market basket." The cost of this basket is then compared over time. This results in a price index defined as the cost of the market basket today as a percentage of the cost of that identical basket was last year.

Consumer Price Index (CPI) is a measure of price changes in consumer goods and services such as gasoline, food, clothing and automobiles. The CPI measures price change from the perspective of the purchaser. U.S. CPI data can be found at the Bureau of Labor Statistics.

Producer Price Indexes (PPI) is a family of indexes that measure the average change over time in selling prices by domestic producers of goods and services. PPIs measure price change from the perspective of the seller. U.S. PPI data can be found at the Bureau of Labor Statistics.

You can think of price indexes as large surveys. Each month, the U.S. Bureau of Labor Statistics contacts thousand of retail stores, service establishments, rental units and doctors' offices to obtain price information on thousands of items used to track and measure Price changes in the CPI (Consumer Price Index). They record the prices of about 80,000 items each month, which represent a scientifically selected sample of the prices paid by consumers for the goods and Services purchased. In the long run, the various PPIs and the CPI show a similar rate of inflation. This is not the Case in the short run, as PPIs often increase before the CPI. In general, investors follow the CPI more than the PPIs

average inflation	inflation	average inflation	inflation
CPI United States 2016	1.26 %	CPI United States 2006	3.24%
CPI United States 2015	0.12 %	CPI United States 2005	3.29%
CPI United States 2014	1.62 %	CPI United States 2004	2.68%
CPI United States 2013	1.47 %	CPI United States 2003	2,27%

Inflation means that the dollar is worth less each year by 1 to 2 percent. This implies that the principal value of the Bond purchased for $1,000 after 5 years is only worth

$1,000 (1 - 5 × Inflation Rate) = $1,000 1 - 5 × 1%) = 95% × $1,000 = $950.

The seller receives $1,000 but the $1,000 only buys $950 worth of goods.

Here are two concrete examples:

One 8 oz cup of coffee in 1940 costs $0.05
One 8 oz cup of coffee in the Hofstra Student Cafeteria costs $1.36
One 8 oz cup of coffee in the Au Bon Pain Coffee Shop costa $1.96

Rate of Inflation for coffee = $1.36/$0.05 = $(1 + R)^{2018 - 1940}$

$27.2 = (1 + R)^{78}$ R = 3.315% for Hofstra

UNTAUGHT MATH   27

For Au Bob Pain $39.2 = (1 + R)^{2018-1940}$ $R = 4.81\%$

If the inflation rate is 1.5%

$20,000 10 years ago is worth less today and $= \$20.000/(1.015)^{10}$ = \$17,233.

Chapter 12.0 What is π?

π is the ratio of circumference to diameter of a circle. Ancient people knew it as 3:1. The approximate value as taught in school was $22/7 = 3.142857143...$ as compared with the true value of $\pi = 3.141592653589793$, etc. The difference is 0.04%.

Chapter 12.1 Euler's Method Calculating π

$= 1 + 1/2^2 + 1/3^2 + 1/4^2 + 1/5^2 + 1/6^2$ etc.....1,000 terms result in 3.1406

Chapter 12.2 Walls Method Calculating π

$$\pi/2 = \frac{2^2 \times 4^2 \times 6^2 \times 8^2 \times 10^2...\text{etc.}}{3^2 \times 5^2 \times 7^2 \times 9^2 \times 11^2...\text{etc.}}$$

π appears in math in equations that involve cycles, repetitive motion, and orbits.

John Walls was the mathematician who assigned two horizontal zeros to represent infinity.

Chapter 12.3 Rabbi Elijah of Vilna's method of computing π (1720-1797)

Computed π from the Bible: Kings I:23 & Chronicles 4:2 as follows:

King Solomon built a ritual bath 10 cubits in diameter and 30 cubits in circumference. The word Circumference in Hebrew is spelled differently in the Biblical Book of Kings and the Book of Chronicles.

Rabbi Elijah of Vilna (Vilna Gaon) was asked why?

In Hebrew Circumference is spelled as two letters in II Chronicles; Kuf followed by a vov. In Kings the same word is spelled Kuf followed by a vov followed by a hey.

The last letter is silent and both words are pronounced as kav. Each Hebrew letter in the alphabet is assigned a number; a = 1, b = 2, g = 3, etc. (aleph, bet, gimel, etc.)

Kuf = 100, vov = 6 & hey = 5. In Chronicles the value of the word circumference = 106. In Kings the value of the word circumference = 100 + 6 + 5 = 111. He computed π as follows:

$(30/10) \times (111/106) = 3.141509434$

$$\pi \text{ (true)} = 3.141592653$$

Percentage error = {3.141592653 -3.141509434}/ 3.141592653 = 0.00264%

Chapter 12.4 Archimedes Method to Calculate π.

Archimedes circumscribed a polygon around a circle of radius = 1.00 and also inscribed a circle inside a polygon with the same number of sides. The areas of both polygons were then added and divided by 2. Since the radius = 1. The averaged area is set equal to π.

A circle is a polygon with an infinite number of sides.

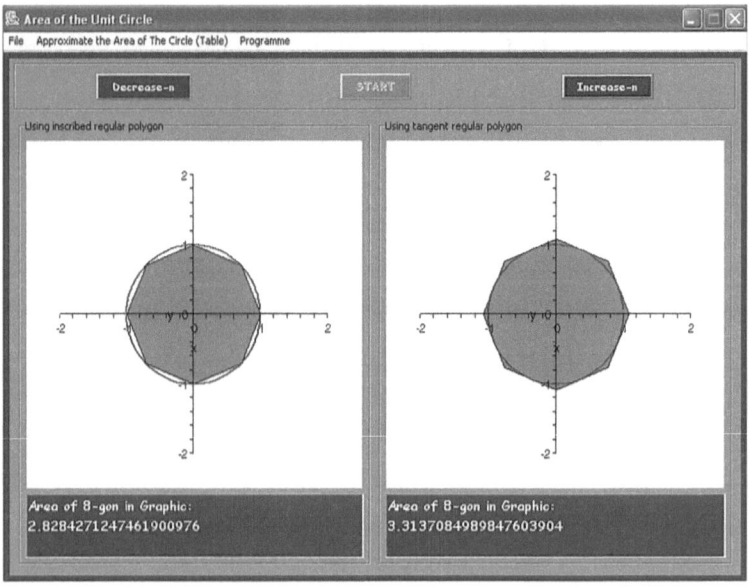

Inside polygon area   Outside polygon area: Average area (approximation of π

6 sides 2.598	3.484	3.041
8 sides 2.828	3.313	3.070
10 sides 2.938	3.249	3.093
16 sides 3.061	3.182	3.121
20 sides 3.089	3.167	3.128
40 sides 3.128	3.148	3.138
100 sides 3.138	3.142	3.140

Inside Area = N Cos {(N-2)(180/2N)} Sin{(N-2)(180/2N)}

Outside Area = N/Tan {(N-2)(180/2N)}

For what value of N will Outside Area = Inside Area?

Setting Areas = to each other results in N = N - 2, Therefore N = Infinity

Chapter 12.5 Number of Integers in to test for repetition of integers

If the integers repeat, then π is not a transcendental number because a repeating decimal can be expressed as a ratio of two numbers. The value of π was calculated for 10^{13} integers and no repeating pattern was observed. π is a transcendental number.

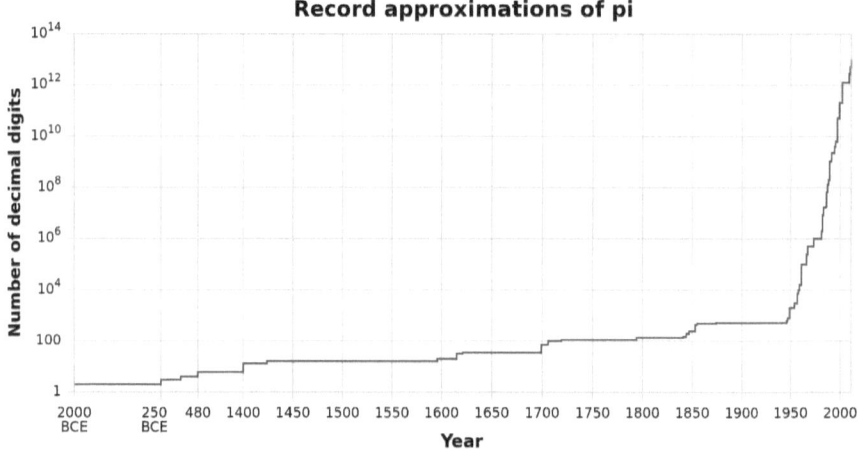

Number of integers in π showing that the integers do not repeat: π is truly a transcendental number.

Chapter 13.0 Imaginary Numbers

A basic imaginary number satisfies equation;

$$X^2 + 1 = 0$$

Solution: x = +i and - i $i = \sqrt{-1}$

i is defined to be an imaginary number.

ASMAD can be used with imaginary numbers

Chapter 13.1 Complex Numbers contain real and imaginary numbers

$$X^3 + 1 = 0$$

Three Roots of equation are:

$X = -1$ $X = \{1 - i\sqrt{3}\}/2$,$X = \{1 + i\sqrt{3}\}/2$

Chapter 13.2 Complex conjugate numbers

Factor: $(X^2 - Y^2) = (X + Y)(X - Y)$

Factor: $(X^2 + Y^2) = (X + iY)(X - iY)$ These factors are called complex conjugates.

Complex numbers are of the form: a + bi and a - bi

Chapter 14.0 Solid geometry

Conic Sections are obtained by slicing a cone in several ways. Conic sections thus obtained are: parabolas, circles, ellipses, hyperbolas and straight lines.

The equations that define each conic section are as follows: A, B, & C are constants.

Parabola: $Y = AX^2 + BX + C$

Circle: $X^2 + Y^2 = R^2$

Ellipse: $X^2/A + Y^2/B = 1$

Hyperbola: $X^2/A - Y^2/B = 1$ or $XY = +/- C$

Straight line: $Y = AX + B$ when the cutting plane is parallel to both edges

Chapter 14.1 Volume of Solids

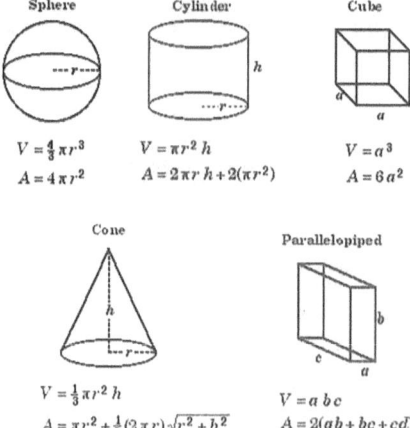

Sphere
$V = \frac{4}{3}\pi r^3$
$A = 4\pi r^2$

Cylinder
$V = \pi r^2 h$
$A = 2\pi r h + 2(\pi r^2)$

Cube
$V = a^3$
$A = 6a^2$

Cone
$V = \frac{1}{3}\pi r^2 h$
$A = \pi r^2 + \frac{1}{2}(2\pi r)\sqrt{r^2 + h^2}$

Parallelopiped
$V = a b c$
$A = 2(ab + bc + cd)$

Volume of a pyramid whose base has 4 equal sides = 1/3 Area of Base × height

Volume of a regular tetrahedron (equal sides) = (Side)³/ (6 $\sqrt{2}$)

Chapter 14.2 Dihedral Angles

When the two intersecting planes are described in terms of Cartesian coordinates by the two equations the dihedral angle Q is given by:

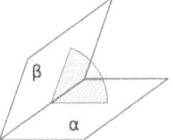

$a_1X + b_1Y + c_1Z + d_1 = 0$

$a_2X + b_2Y + c_2Z + d_2 = 0$

Let Q = angle between planes, d1 and d2 are constants.

$\text{Cos } Q = \dfrac{a_1a_2 + b_1b_2 + c_1c_2}{[a_1^2 + b_1^2 + c_1^2]^{1/2} [a_2^2 + b_2^2 + c_2^2]^{1/2}}$

Chapter 15.0 Vector addition

Vectors have direction and magnitude. They are useful in describing force and motion. In the X-Y Plane below we see two vectors one with a direction of 30 degrees from the horizontal axis with a magnitude of 5 and the other with a direction of 60 degrees with a magnitude of 6. These vectors may be added by combining the force components in the X and the Y directions. In the X direction The components are X = 5 Cos 30 deg. and X = 6 Cos 60 deg. In the Y direction we have components Y = 5 Sin 30 deg. and 6 Sin 60 deg. Therefore X = 5 Cos 30 + 6 Cos 60 = 7.33

Therefore, Y = 5 Sin 30 + 6 Sin 60 = 7.69

The magnitude of the new vector (or sum) = $\sqrt{7.33^2 + 7.69^2}$ = 10.62

The angle of the resultant = ARC TAN (7.69/7.63) = 45.2 degrees

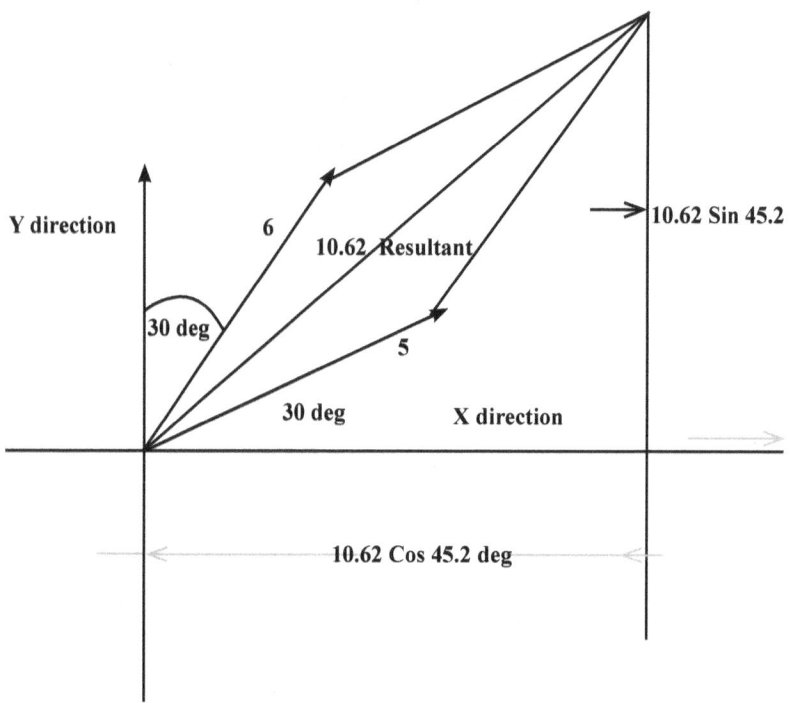

Two vectors may be multiplied (called a dot product) or a cross product. The dot product is defined as follows:

Vectors X.Y are multiplied as a dot product by multiplying the resultant of both vectors by the cosine of the angle between them.

Chapter 15.1 Multiplying Two Vectors (cross product)

The vector cross product is defined as follows:

Vectors X.Y are multiplied as a cross product by multiplying the resultant of both vectors by the Sin of the angle between them.

$X \times Y$ is a cross product = 5.0 × 6.0 Sin 30 deg. (angle between X and Y)

$X \times Y = 15.0$ Result is a vector perpendicular to the plane of X and Y. The cross product is a rotation of X onto Y or Y onto X; so the direction of rotation matters so that $X \times Y = - Y \times X$

Chapter 15.2 Three Dimensional Vectors

For three vectors, A, B, & C (three dimensions)

The Resultant vector from the origin is the geometric sum of V_x, V_y & V_z.

V (Resultant) = V(x) + V(y) + V(z] in vector representation or $V^2 = X^2 + Y^2 + Z^2$

ASMAD is possible by operating separately on the projected vectors in each plane. ASMAD has to be performed separately for each X, Y, & Z component if there are multiple vectors.

Vectors represent 3 dimensional real numbers. a, b ,& c, are projections on three planes each of which is two dimensional XY, XZ, & YZ planes.

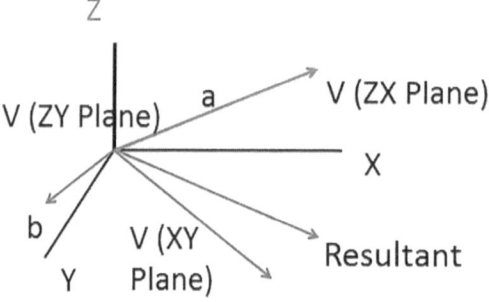

Chapter 15.3 Resultants of vectors

Two Dimensional vectors - Cartesian Coordinates

$R^2 = (X_2 - X_1)^2 + (Y_2 - Y_1)^2$

Three Dimensional vectors - Cartesian Coordinates

$R^2 = (X_2 - X_1)^2 + (Y_2 - Y_1)^2 + (Z_2 - Z_1)^2$

Four Dimensional vectors - (Cartesian Coordinates for space –time)

$R^2 = (X_2 - X_1)^2 + (Y_2 - Y_1)^2 + (Z_2 - Z_1)^2 + (i\ C\ [T_2 - T_1])^2$

$i = \sqrt{-1}$ C = velocity of light, $T_2 - T_1$ = Event Time Duration

Chapter 16.0 Complex number raised to a power

Compute $(1 + i)^n$. In this example, let n = 8. This is a long method.

$(1 + i)^2 = (1 + i)(1 + i) = 1 + 2i + i^2 - 1 + 2i - 1 = 2i$

$(1 + 1)^3 = (1 + i)^2 (1 + i) = 2i (1 + i) = 2i - 2 = 2(i - 1)$

$(1 + 1)^4 = (1 + i)^2 (1 + i)^2 = (-2i)(-2i) = -4$

$(1 + 1)^5 = (1 + i)(-4) = -4 - 4i$

$(1 + 1)^6 = (2(i - 1))(2(i - 1)) \; 4(-2i) = -8i$

$(1 + 1)^7 = (-8i)(1 + i) = -8 - 8i$

$(1 + 1)^8 = (-4)(-4) = 16$

Chapter 16.1 Graphical method to compute $(1 + i)^m$

Complex numbers may be represented in the complex plane where X-axis (horizontal) represents the set of integers and the Y-axis (vertical) represents the set of imaginary numbers. From de Moivre's theorem w have

$$e^{i\theta} = \cos\theta + i \sin\theta$$

$$\{e^{i\theta}\}^n = \{\cos\theta + i \sin\theta\}^n$$

$$(1 + i)^8 = \{\sqrt{2}\}^8 \; \{\cos(8)(45) + i \sin(8)(45) \; |\}$$

$$= 16 \{\cos 360 + i \sin 360\}$$

$(1 + i)^8 = 16$ Both legs of the right triangle in the figure below equals 1.0 for which the resultant equals 16 and 8 × 45 = 360.

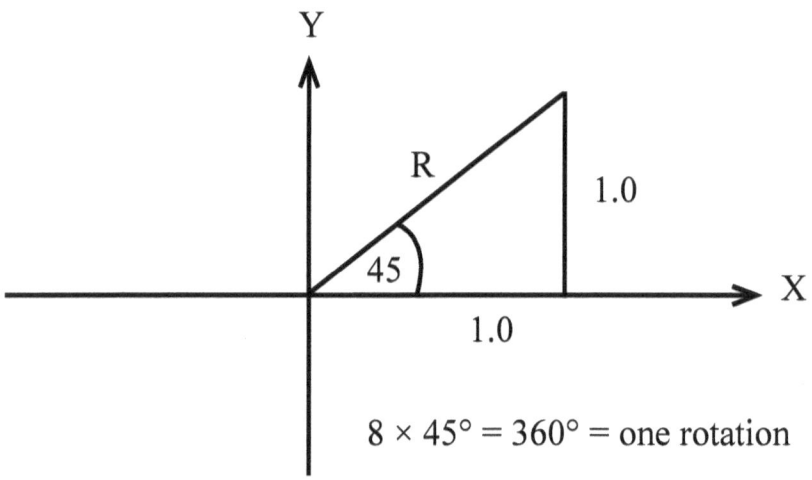

$8 \times 45° = 360° = $ one rotation

$(\sqrt{2})^8 = 16$
$R^2 = 1^2 + 1^2 = 2$

Chapter 17.0 Tensors

The tensor is an advanced vector and consists of nine components that completely define the state of a material object or space itself. An example is the stress tensor. At a point inside a material the tensor defines the forces in or on the material. The tensor relates a unit-length direction vector to the stress vector that results from, torsion, (twisting), tension/compression, or shear (parallel to the surface).

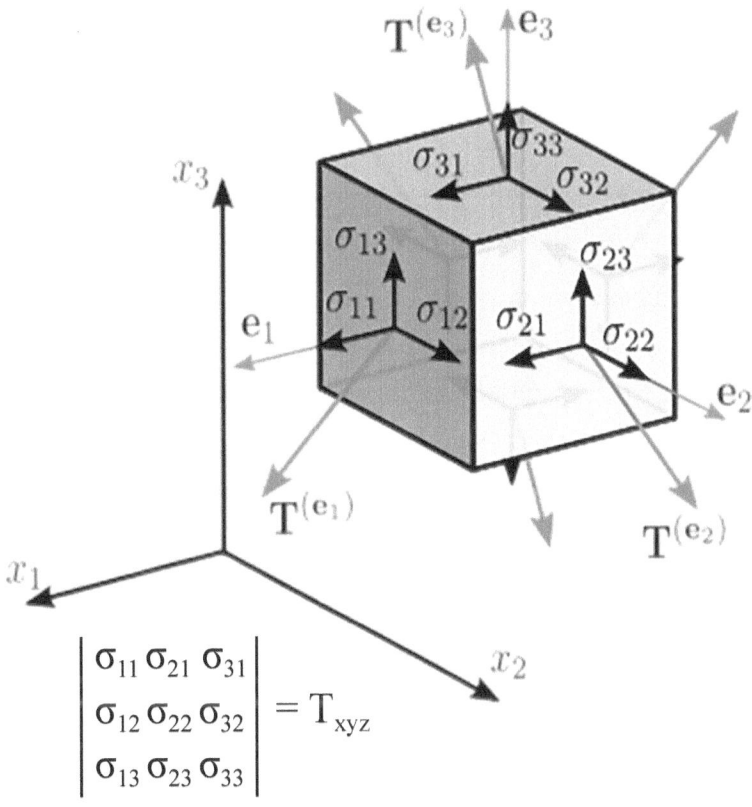

$$\begin{vmatrix} \sigma_{11} & \sigma_{21} & \sigma_{31} \\ \sigma_{12} & \sigma_{22} & \sigma_{32} \\ \sigma_{13} & \sigma_{23} & \sigma_{33} \end{vmatrix} = T_{xyz}$$

σ_{11}, σ_{12} & σ_{13} are the components of tension or compression torsion and shear of the vectors in

the X_1, X_2, & X_3 directions in lbs./inch² on one plane. There are three sets of three vectors

because there are three planes and the 9 stress components are represented by a 3×3 matrix.

T_{xyz} is the stress matrix. There is also a strain matrix = S_{xyz} such that. $T_{xyz} = S_{xyz} \times K_{xyz}$

Stress = Strain X a constant (may not be a constant if the stress or strain varies with time)

Chapter 17.1 Viscoelasticity Definition

Materials whose response to the stress tensor varies with time are called viscoelastic. Examples are molasses, tar, honey and soft plastics. Such materials do not return to their initial position when stress is removed. The tensor that describes viscoelasticity has four dimensions (length, width, height, and time. There are 16 terms in the viscoelastic stress and strain tensors. tensor. The matrices have to be multiplied To solve the equations which may not be linear.

Chapter 18.0 Euler Series for calculating "e" and proving DeMoivre's Theorem.

The great mathematician Leonhard Euler discovered the number e and calculated its value to 23 decimal places. It is often called Euler's number and, like, is a transcendental number (this means it is not the root of any algebraic equation with integer coefficients).

i = imaginary number

$i = \sqrt{(-1)}$; $i^2 = -1$, $i^3 = -i$, $i^4 = +1$

$e^x = 1 + x + x^2/2! + X^3/3! + X^4/4! + X^5/5! + X^6/6!$ ...... etc....

$e^{ix} = 1 + iX - x^2/2! - i X^3/3! + X^4/4! +i X^5/5! - X^6/6!$

The series to calculate Sin's and Cosines are as follows:

$\cos x = 1 \quad - X^2/2! \quad + X^4/4! \quad - X^6/6! \ldots\ldots$etc.

$i \sin X = iX \quad -i X^3/3! \quad +i X^5/5! \quad -i X^7/7!\ldots\}\ldots$etc.

Adding all terms: $e^{ix} = \cos x + i \sin X$ This is DeMoivre's theorem X is the angle in radians. 2π radians = 360 degrees. 1 radian = 57.3 degrees.

If $x = \pi$ radians then,

$e^{i\pi} = -1$ or $e^{i\pi} + 1 = 0$ which is one of the most important equations in all of mathematics. The equation contains the six major parameters in all of mathematics;

$$e, i, +, 1, \pi \text{ and } 0.$$

DeMoivre's Theorem enables us to calculate the roots of complex numbers as shown in Chapter 16.1

Chapter 19.0 Golden Ratio, Length/Width of Rectangle

Starting with a triangle whose sides have a ratio of 2:1, and the hypotenuse = $\sqrt{5}$, we rotate the hypotenuse 90 degrees. = $\sqrt{5}$ - 1. We rotate the hypotenuse 90 degrees from the vertex on top to the edge on the bottom such that the width of the rectangle = $\sqrt{5}$ - 1. The length = 2. The length to width ratio = 1.618. The Golden Rectangle, was proposed by the Greek philosophers, as the most esthetically attractive rectangle. Their buildings incorporated the

Golden Rectangle in their facades and the design of the Parthenon incorporated the ideas of the Golden Ratio.

Golden Rectangle - as proposed by the Greek philosophers as the most aesthetically attractive rectangle.

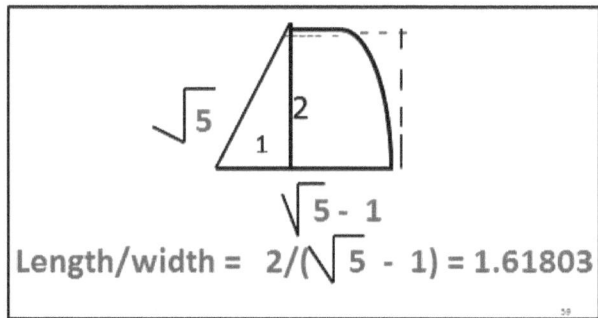

Johannes Kepler recognized that the base area of the Khufu Pyramid in Egypt was 52,000 meters^2 and the combined area of the four sides is 85,647 meters^2. The ratio = 1 + 52,900 / 85,647 = 1.618 which the early philosophers called the "Divine Proportion." Leonardo da Vinci observed that the ratio of the height divided by the distance of the naval from the floor is the divine ratio. For me, 64 inches divided by 38 inches = 1.684, close to the divine ratio.

Chapter 20.0 Leonardo Pisano Fibonacci

Leonardo Fibonocci 1170–1240

Fibonacci Numbers

1,1,2,3.5, 8,13, 21, 34, 55, 89, 144, 233, 377, 610, 987, etc... are obtained by starting with "1" and adding the next number:

1 + 1 = 2, 1 + 2 = 3, 2 + 3 = 5, 3 + 5 = 8, 5 + 8 = 13, 8 + 13 = 21. etc

The numbers represent growth patterns of pine cones, sunflower seeds, sea and shell spirals. The division of successive Fibonacci Numbers by the previous number converges to the Golden Ratio of the length to to Width (L/W) ratio of 1.61803. Here are examples:

3/2 = 1.500

5/3 = 1.666

8/5 = 1.600

13/8 = 1.625

21/13 = 1.615

34/21 = 1.619

$55/34 = 1.617$

$89/55 = 1.618$

$144/89 = 1.617$

$233/144 = 1.618$

$377/233 = 1.618$

$610/377 = 1.618$

$987/610 = 1.618$

No reason has been found to explain and correlate the division of two sequential Fibonacci

Numbers with the Golden ratio = 1.618 for a rectangle.

$2/(\sqrt{5}-1) = 1.618 = (1 + \sqrt{5})/2 = 1.618$

Here is an algorithm for computing every other Fibonacci Number:

1, 1, 2, 3, 5, 8, 13, 21, 34, 55, 89, 144, 233, 377, 610, 987, 1597, 2584, 4181….

$1^2 + 1^2 = 2$

$1^2 + 2^2 = 5$

$2^2 + 3^2 = 13$

$3^2 + 5^2 = 34$

$5^2 + 8^2 = 89$

$8^2 + 13^2 = 233$

$13^2 + 21^2 = 610$

$21^2 + 34^2 = 1597$

$34^2 + 55^2 = 4181$

Chapter 20.1 Srinivasa Ramanujan

In the movie, "The Man Who Knew Infinity" George Hardy (mathematician at Trinity College) escorted Ramanujan to a cab to take him to the airport in a return trip to India. Prof. Hardy remarked that the number on the cab was 1729 and this was not a significant number. Ramanjujan disagreed and quickly said that 1729 is the sum of two pair of cubes. $1729 = 12^3 + 1^3$ and also $1729 = 10^3 + 9^3$.

Ramanujan was a math genius who, unfortunately died at age 32 of pneumonia he contracted while living in London. He was not used to living in such a cold climate. He was born in 1887 and died in 1920 during of the time of the Spanish Influenza epidemic. He discovered an algorithm whose numbers converge to the Golden Ratio of 1.618. The corresponding values are shown in the vertical column.

$$1 + \cfrac{1}{1 + \cfrac{1}{1 + \cfrac{1}{1 + \cfrac{1}{1 + \cfrac{1}{1 + \cfrac{1}{1 + \cfrac{1}{1 + \cfrac{1}{1 + 1}}}}}}}} \quad \text{etc}$$

1.50000
1.66666
1.60000
1.62500
1.61538
1.61904
1.61764
1.61818

Chapter 20.2 Golden Ratio for a pentagon

Diagonal of a Pentagon = Golden Ratio if each side = 1.0
Use Law of Cosines

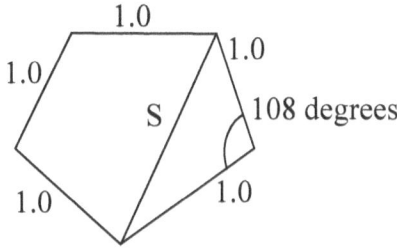

$S^2 = (1)(1) + (1)(1) - 2(1)(1) \cos 108$

$S = 1.61802$

Chapter 20.3 Fibonacci Patterns

Tree Branches grow in Fibonacci Pattern

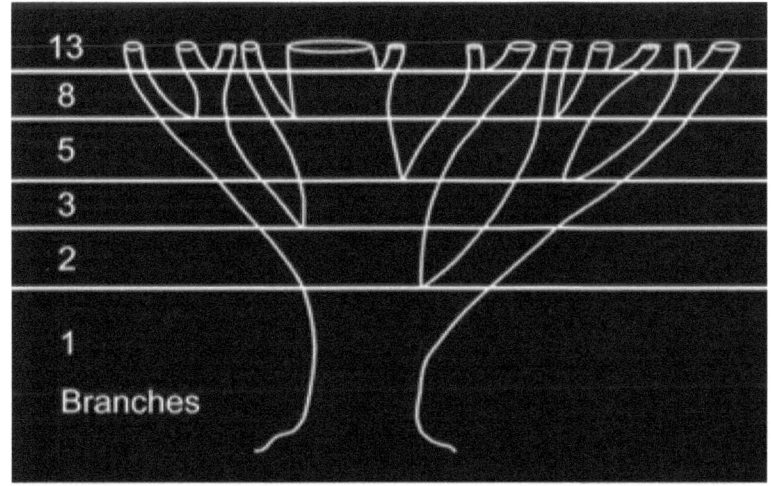

Fibonacci Spiral

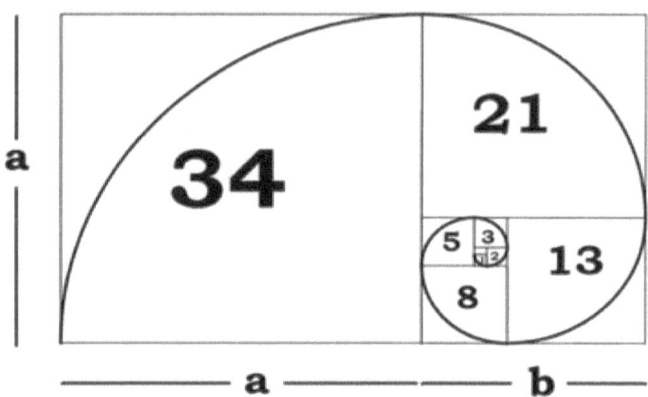

Logarithmic Spiral based on golden ratio

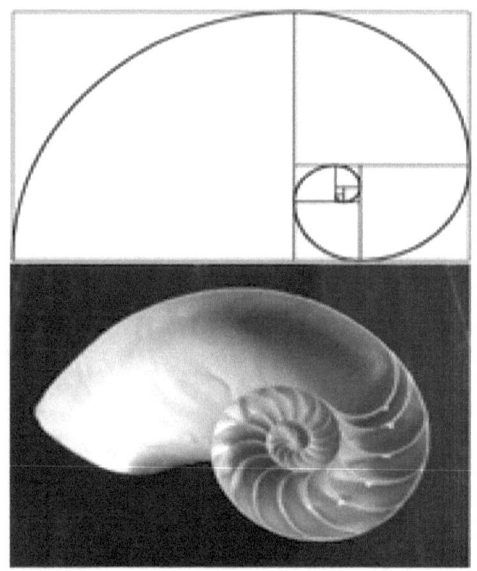

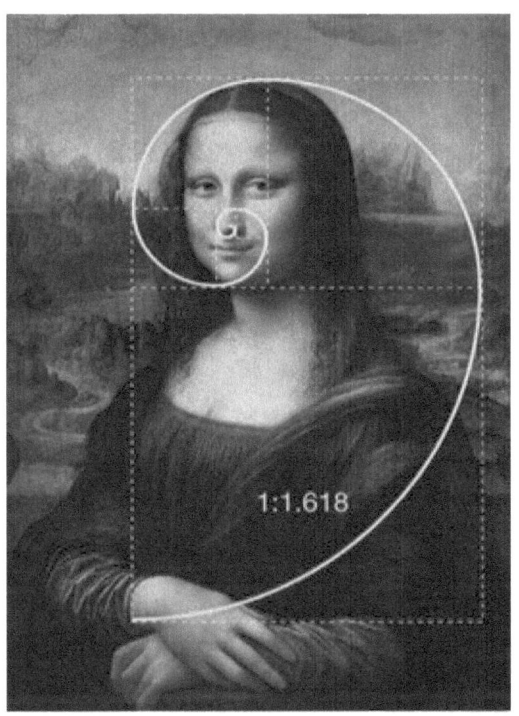

Chapter 20.4 Quadratic Equation Solution & Golden Ratio

$$X^2 - X - 1 = 0$$

X = 1.618 and X = -0.618, X is the only number when squared and then subtracted from the squared number: 1.618 – 0.618 = 1.000. Also, 1/(-0.618) = -1.618. Strange!

Chapter 20.5 Dr. Robert Weller's Axiom

n	n^2	$(n+1)^2$	Difference = $2n^2 - 1$
1	1	4	4 - 1 = 3
2	4	9	9 - 4 = 5
3	9	16	16 - 9 = 7
4	16	25	25 - 16 = 9
5	25	36	36 - 25 = 11
6	36	49	49 - 36 = 13
7	49	64	64 - 49 = 15
8	64	81	81 - 64 = 17
9	81	100	100 - 81 = 19
10	100	121	121 - 100 = 21

Etc ......

In the column on the right, the difference always = 2.

The General Expression is: $(N_i + 1)2 - (N_i)2 = 2N_i + 1$

Example: 1) $7^2 - 6^2 = 2 \times 6 + 1$

Example: 2) $30^2 - 29^2 = 2 \times 29 + 1$

After algebraic substitution: $N_i + 1 + N^i = 2N^i + 1$

Example: 3) $8 + 7 = 2 \times 7 + 1$

Example: 4) $256 + 255 = 2 \times 255 + 1$

Dr. Robert Weller is a retired dentist with an interest in numeric.

Chapter 20.6 Zeno's paradox or you can't get there from here

I want to go from point A to point B by taking steps equal to ½ of the remaining distance between point and point B.

A..B

The distance between points A and point B = L

First Step = ½ L

Second Step = ½ of ½ of L = ¼ L

Third Step = ½ of ¼ L = ⅛ L.

Fourth Step = ½ of ⅛ L = 1/16 L

So we have an infinite series as follows:

½ + ¼ + ⅛ + 1/16 + 1/32 + 1/64 + 1/128 ......

Will all these steps converge to point B on L?

Step no.	Distance traveled per step	total distance traveled
1	0.5000	0.500000
2	0.2500	0.750000
3	0.1250	0.875000
4	0.0625	0.937500
5	0.03125	0.968750
6	0.01562	0.984375
7	0.00781	0.992185

……etc. So you can see that I need to make an infinite number of steps to reach Point B on Line L. Since I can never make an infinite number of steps, reaching Point B is impossible. We can say that the series converges to point B only after an infinite number of steps. Since we can't walk an infinite number of steps, then we can't arrive at Point B.

Chapter 20.7 Infinite Series Convergence

The series $\frac{1}{2} + \frac{1}{4} + \frac{1}{8} + \frac{1}{16} + \frac{1}{32} + \frac{1}{64} + \frac{1}{128}$ etc…is called an infinite series. There is a simple test for convergence called the ratio test. If the ratio of the forward number divided by the back number< 1, then the series converges at infinity.

Example 1) 1/8/1/4 = 0.5. 1/64/1/32 = 0.5, therefore this series converges.

Chapter 20.8 Casting out Nines – First Method

Select any number with more than two integers. Subtract the sum of the integers from the number selected and the answer is always divisible by nine.

Example 1) 5316 The sum of the integers is: 5 + 3 + 1 + 6 = 15.

Subtract 15 from 5316 = 5301. 5301/9 = 589

Then 5316 = 9 × 589 + 15

Example 2) 297. The sum of the integers = 2 + 9 + 7 = 18, 1 + 8 = 9.

287 - 9 = 288, 288/9 = 32

Chapter 20.9 Casting out nines- Second Method

Use same number 5316. Sum of integers = 15. Subtract 15 from 5316, answer = 5301. 5301/9 589. Therefore 5316 = 9 × 589 + 15. Instead of subtracting 15, we can add 3 to 5316 because 15 + 3 = 18 and 18 is a multiple of 9. So we get 5316 + 3 = 5319; 5319/9 = 591. Therefore 5316 = 591 × 9 - 3.

Chapter 30.0 Two Digit Reversal Math

Pick any two digit number. Subtract the digit reversal. Add the digit reversal to the answer. The final answer is always = 99. Here are examples:

76 - 67 = 09, 90 + 09 = 99

53 - 35 = 18, 81 + 18 = 99

51 - 15 = 36, 63 + 36 = 99

92 - 29 = 63, 36 + 63 = 99

72 - 27 = 45, 45 + 54 = 99

Chapter 30.1 Three Digit Reversal Math

Pick any three digit number. Subtract the digit reversal. Add the digit reversal to the answer.

The final answer is always 1089. Here are examples:

723 - 327 = 396. 396 + 693 = 1089\
953 - 359 = 594. 594 + 495 = 1089
634 - 436 = 198. 198 + 891 = 1089
381 - 183 = 198. 198 + 891 + 1089

Chapter 30.2 Five regular polyhedrons

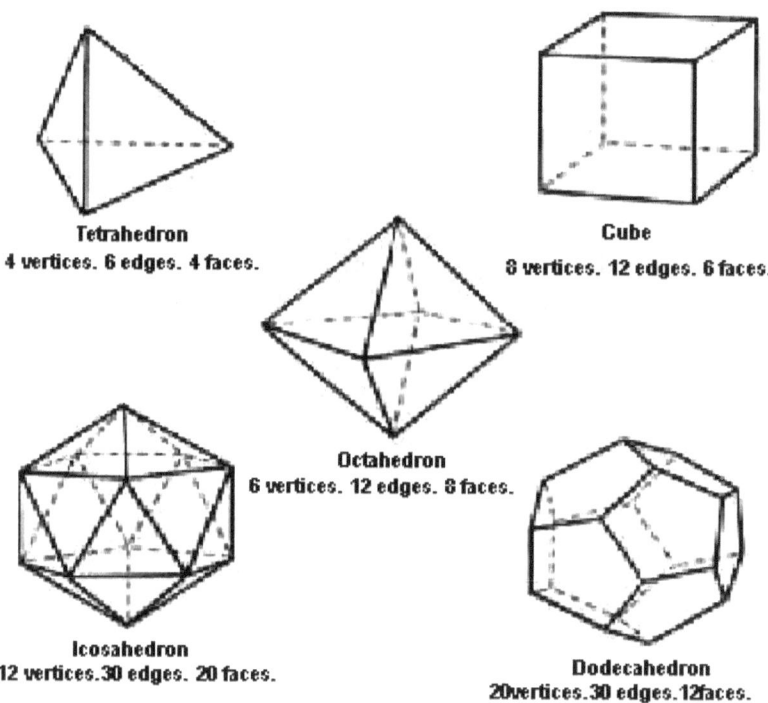

Tetrahedron
4 vertices. 6 edges. 4 faces.

Cube
8 vertices. 12 edges. 6 faces.

Octahedron
6 vertices. 12 edges. 8 faces.

Icosahedron
12 vertices. 30 edges. 20 faces.

Dodecahedron
20 vertices. 30 edges. 12 faces.

The dihedral angles (angles between each face) for each polygon are equal.

Polyhedron Type	Vertices	Faces	Edges	V + F - E
Tetrahedron	4	4	6	2
Cube	8	6	12	2
Octagon	6	8	12	2
Icosahedron	12	20	30	2
Dodecahedron	20	12	30	2

Therefore V + F = E + 2 This is known as Euler's Formula. It works for most solids.

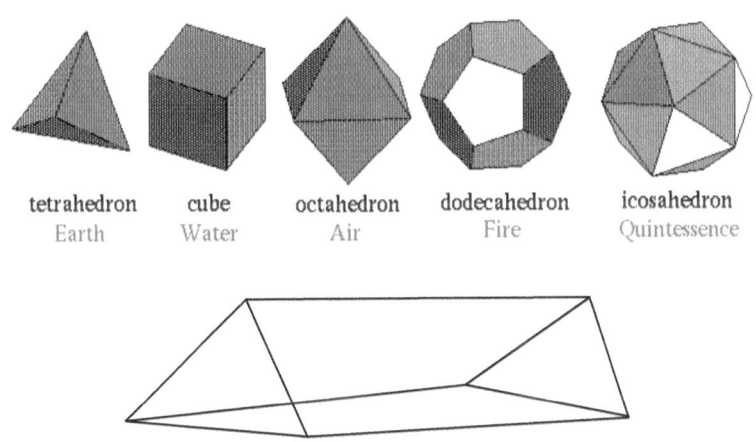

The five Platonic solids

tetrahedron — Earth
cube — Water
octahedron — Air
dodecahedron — Fire
icosahedron — Quintessence

Sum of vertices = 6. Sum of the edges = 9. Sum of the faces = 5.

Therefore 6 + 5 – 9 = 2

Type of solid	length or radius	volume	surface area	surface area/ volume
Sphere	radius = 1.0	$4\pi/3$	4π	3.0
Hemisphere	radius = 1.0	$2\pi/3$	$2\pi/3 + \pi/2$	3.75
Cube	side = 1.0	1.0	6.0	6.0

UNTAUGHT MATH

Tetrahedron	side = 1.0	0.118	1.732	14.68
Octahedron	side = 1.0	0.471	3.464	7.35
Dodecahedron	side = 1.0	7.663	20.646	2.694\
Icosahedron	side = 1.0	2.182	8.660	3.969

The dodecahedron uses the least surface area to enclose a fixed volume.

Chapter 30.3 Regular Polygons (side or radius = 1.0)

Type of polygon	area	perimeter	perimeter/area
Square	1.0	4.0	4.0
Circle	$\pi 2$	2π	2.0
Hexagon	2.599	6.0	2.308
Isosceles triangle	0.433	3.0	6.928
Octagon	3.6952	8.0	2.1649
Semi-circle	$\pi/2$	$\pi + 2$	3.273

The circle uses the smallest perimeter to enclose a fixed area.

Chapter 40.0 Binary Numbers Gottfried Wilhelm Leibnitz

Binary Numbers were conceived by Gottfried Wilhelm Leibniz (1646 – 1716) and consist of two numbers, 0 and 1. Every number can be represented by a zero and one

Base 10 numbers	Base 2 Numbers (Binary)
1	0
2	10
3	11
4	100
5	101
6	110
7	111
8	1000
9	1001
10	1010
11	11011
12	10100

An in-depth explanation is presented in the cryptography (Chapter 60.0) Section in this book

Chapter 41.0 Prime Numbers and Cryptography

Cryptography is essential to provide for privacy for emails, access to bank accounts, etc. Prime Numbers are essential in Cryptography.

The fundamental Theorem of Arithmetic is that every integer is either a prime number or the product of prime numbers. A prime number is only divisible by number 1 or by itself. Examples of prime numbers: 1, 5, 17, 23, 51

Number 12 has many factors therefore it is not a prime number.

The number 100 has many factors so it is not a prime number.

Every even number is the sum of two prime numbers. Here are some examples:

24 = 23 + 1, 40 = 37 + 3, 100 = 97 + 3

Some Questions about Prime Numbers

- Of what use are prime numbers?
- Who selects the prime numbers used for Cryptography.?
- Can they all be listed as a table for code-breakers to use?
- Can trying all the possible keys break Pretty Good Privacy (PGP)?
- How do we know a large number is a prime number?
- How can we generate large prime numbers?
- What is the largest prime number?
- How many prime numbers are there?

Prime numbers are selected for cryptography because the product of two large prime numbers results in a very large non-prime number that is difficult to factor thereby frustrating a code-breaker or hackers. Mathematicians have been studying factoring for centuries and present methods are not much better than ancient methods. A large computer is needed. How large a product is needed to discourage code breaking? Examples of the Product of large Prime Numbers (Wall Street Journal 6/16/2018)

1, 202, 337, 603 × 8, 486, 003 = 10, 203, 040, 506, 070, 809

995, 465, 061 × 91, 220, 741 = 90, 807, 060, 504, 030, 201

The numbers are too large to use Fermat's Theorem on a hand calculator.

Can two people be assigned the same prime number by the computer's Random Prime Number Generator? Very unlikely! Who selects the prime numbers for cryptography?

A computer selects the two large prime numbers from a random prime number generator and multiplies them to produce part of the public key. The other part of the public key is the exponent.

How many prime numbers are there? Euclid in 300 BC proved that there are an infinite number of prime numbers. His proof is still accepted. There are approximately 10^{151} prime numbers of length 512 bits or less. This number exceeds the number of atoms in the universe. No of Primes approximately = N/ln N. See Table I.

Can all primes be listed as a table for code-breakers?

No! There are too many.

Can the Pretty Good Privacy (PGP) algorithm be broken by trying all the possible keys?

Assume a 128 bit key is used. Assume a billion chips can scan a billion keys per second.

Time required is about a thousand times age of universe.

Table I   Distribution of Prime Numbers

Formula of Karl Friedrich Gauss and Adrien Marie Legendre

Number of primes approximately =
$N/(\operatorname{Ln} N) = N/[(\operatorname{Ln} 10)(\operatorname{Log}_{10} N)]$

$\operatorname{Ln}(10) = 2.302585093$

Table I

Range	Estimated No. of prime numbers	actual
1- 10	4	4
1- 100	21	25
1- 1,000	145	168
1- 10,000	1,086	1,229
1- 100,000	8,686 = (100,000)/[(2.302585093)(5)] est	
1- 1,000,000	72,382	78,498
1- 10,000,000	620,420	
1- 100,000,000	5,428,681	
1- 1,000,000,000	48, 254,942	approx. = $2^{25.524}$
1-10,000,000,000	434,294,481	1- 10^{20} = 2,171,472,410 ×10^9

= 2,220,819,602,560,918,840

In 2017 the largest prime number reported in the Wall Street Journal is:

$2^{77,232,917}- 1$; it has 23,249,425 integers, a mega-prime.

The formula is not perfect for predicting the number of prime numbers within a certain range.

How do we know a large number is a prime number?

We can use Fermat's test.

Chapter 41.1 Fermat's Test for Prime Numbers

Fermat's test: If 2^p {modulo p} = 2, then p is a prime number.

Example 1) p = proposed prime = 19

2^{19} {modulo 19} = 524288/19 = 27594 + Remainder of 2, therefore 19 is a prime number.

Example 2) p = proposed prime = 18

2^{18} {modulo 18} = 262144/18 = 14563 + Remainder of 10

Therefore 18 is not a prime number and that is an expected conclusion.

Chapter 41.2 Algorithm for generating prime numbers.

The Mersenne algorithm produces non-primes. There is no known single polynomial that produces all prime numbers and only prime numbers.

Table II   Mersenne prime = $2^p - 1$;

p = prime number

Mersenne   Prime   Numbers

Prime No.	Base 10	Binary or base 2	
1	1	1	
2	3	11	
3	7	111	
5	31	11111	
7	127	1111111	
11	2,047	11111111111 = (23)(89) Fermat's Theorem fails for 2047.	
13	8,191	1111111111111	
17	131,071	11111111111111111	
19	524,287	1111111111111111111	
23	8,388,607	11111111111111111111111	
29	536,870,911	11111111111111111111111111111	Is this Prime?

Yes, Use Fermat's Test: 2^{29} = 536870912,

Does 2^p (modulo p) = 2? Remainder = 536870912/29 = (18512790.0691-18512790) (29) = 2.00

Yes it is a Prime number.

Extremely large products of primes are difficult to factor.

Example: 127×8191 = 1,040,257 = 11101111110000001

We need to use the product of two large Prime numbers for cryptographic encryption.

The largest Prime Number discovered in 2017 is:

$2^{77,232,917} - 1$; it has 23,249,425 integers, it is a mega-prime.

How large a product is needed to discourage code-breaking? We are talking about the product of two Prime numbers. The product of two primes needs to be at least 10^{308} for important banking transactions (The Code Book). Mathematicians are working to factor the products of large primes.

In 1977, a product of two primes having 129 digits presented as a challenge, approx. (10^{129}), was factored in 1994.

Can two people have the same prime number? PGP provides for selection of keys having 512, 2048, or 4096 bits. There are 10^{151} prime numbers from which to choose whose length is less than or equal to 512 bits. The probability of two people having the same prime number, selected from a random process, is less than the probability of all the molecules in a room suddenly moving to one corner depriving the occupants of oxygen.

Importance of Prime Numbers

The product of two large prime numbers is used to establish cryptographic keys in the RSA algorithm (described later in this book. Large computers and lots of time are required to identify the two prime numbers that constitute the product of two large prime numbers. Round-off numbers used in scientific notation and engineering calculations are not allowed.

Chapter 42.0 Perfect Numbers

A perfect number is defined as having all the factors equal to the number.

For example, the factors of 6 are: 3, 2 & 1. The sum of 3, 2, & 1 = 6.

Another example: The factors of 28 are: 14, 2, 4, 7, & 1. The sum = 28.

Euclid showed that any number N is a perfect number if $N = (2^{m-1})(2^m - 1)$

and if $2^m - 1$ is a prime number.

Example: Let m = 5 (prime number); $2^5 - 1 = 31$ (which is a prime number)

$(2^4) \times (2^5 - 1) = 16 \times 31 = 496$ which is a perfect number for which the factors are: 1, 2, 248, 4, 124, 8, 62, 31, 16 Their sum = 496.

Perfect numbers

6, 28, 496, 8128, 33550336, 8589869056, 137438691328, 2305843008139952128, 2658455991569831744654692615953842176, 191561942608236107294793378084303638130997321548169216, 13164036458569648337239753460458722910223472318386943117783728128, 14474011154664524427946373126085988481573677491474835889066354349131199152128,

"Perfect numbers like perfect men are very rare." Rene Descartes.

No useful applications have emerged from the study of Perfect numbers. However Perfect Numbers have been used to amplify Biblical Theories about the existence of a creator. The first perfect number = 6. The first word in the Hebrew Bible "in the Beginning" has 6 letters implying that creation was perfect.

The second perfect number = 28 and the sum of all the letters in the first sentence of the Hebrew Bible has 28 letters "In the Beginning God created the heavens and the earth."

The third perfect number = 496 and equals the gematria of the words "Tablets of Stone" (Exodus 24:12) that Moses carried down from Mt. Sinai. Hundreds of years elapsed before uses were found for imaginary and binary numbers. The same may be said of Perfect Numbers.

Chapter 43.0 Smith Numbers

Smith Numbers are integers for which the sum of the digits equals the sum of all the prime factors.

Examples $27 = 3 \times 3 \times 3 = 2 + 7 = 9$, $94 = 47 \times 2$; $4 + 7 + 2 = 9 + 4 = 13$;

Other examples of Smith Numbers: 4, 22, 27, 58, 85, 94, 121, 166, 202, 265, 274, 319, 346,

Consecutive Smith Numbers below 10^5

(728, 729), (2964, 2965), (3864, 3865), (4959, 4960), (5935, 5936), (6187, 6188), (9386, 9387), (9633, 9634), (11695, 11696), (13764, 13765), (16536, 16537), (16591, 16592), (20784, 20785), (25428, 25429), (28808, 28809), (29623, 29624), (32696, 32697), (33632, 33633), (35805, 35806), (39585, 39586), (43736, 43737), (44733, 44734), (49027, 49028), (55344, 55345), (56336, 56337), (57663, 57664), (58305, 58306), (62634, 62635), (65912, 65913), (65974, 65975), (66650, 66651), (67067, 67068), (67728, 67729), (69279, 69280), (69835, 69836), (73615, 73616), (73616, 73617), (74168, 74169), (74298, 74299), (76495, 76496), (76911, 76912), (77385, 77386), (78744, 78745), (82488, 82489), (82640, 82641), (83744, 83745), (83928, 83929), (83937, 83938), (84759, 84760), (84882, 84883), (85135, 85136), (87362, 87363), (87855, 87856), (89743, 89744), (89904, 89905), (90228, 90229), (90872, 90873), (91255, 91256), (91364, 91365), (91488, 91489), (93275, 93276), (93471, 93472), (94094, 94095), (94184, 94185), (94584, 94585), (95277, 95278), (95984, 95985), (96151, 96152), (96921, 96922), (97915, 97916), (98022, 98023) and (98900, 98901)

A palindrome is an arrangement of a word or a number for which the first letter or number is the same as the last. The second letter or number is the same as the next to last, etc.

Palindromic Smith Numbers < 10^6

22, 121, 202, 454, 535, 636, 666, 1111, 1881, 3663, 7227, 7447, 9229, 10201, 17271, 22522, 24142, 28182, 33633, 38283, 45054, 45454, 46664, 47074, 50305, 51115, 51315, 54645, 55055, 55955, 72627, 81418, 82628, 83038, 83938, 90409, 95359, 96169, 164461, 173371, 239932, 256652, 262262, 294492, 362263, 373373, 445544, 454454, 505505, 515515, 535535, 545545, 635536, 704407, 717717, 832238, 841148, 864468, 951159, 956659, 974479 and 983389

I know of no useful applications of Smith Numbers

Chapter 44.0 Magic Numbers

Magic Numbers are labeled by chemists to describe the maximum number of electrons in each shell that orbit the nucleus of each atom. Table as follows:

Shell Number	Maximum number of electrons per shell	Formula	Full Electrons	
1	2	2×1^2 or	$1 \times 2 = 2$	He 2
2	8	2×2^2 or	$2 \times 4 = 8$	Ne 10
3	18	2×3^2 or	$3 \times 6 = 18$	Ni 28
4	32	2×4^2 or	$4 \times 8 = 32$	Nd 60
5	50	2×5^2 or	$5 \times 10 = 50$	Ds 110
6	72	2×6^2 or	$6 \times 12 = 72$	None 182
7	98	2×7^2 or	$7 \times 14 = 98$	None 280

The numbers of electrons per shell grow in a mathematical sequence.

Shells 5, 6 & 7 are never completely filled but their capacities are listed.

Further Analysis:

8 - 2 = 6

18 - 8 = 10

32 - 18 = 14

50 - 32 = 18

72 - 50 = 22

98 - 72 = 26

The numbers on the right side grow by 4.

Conclusion: The designer of the world is a mathematician.

Chapter 44.1 Magic Calculations

Select any 3 digit number and then repeat: Divide by 1001: 452452/1001 = 452

Select any 4 digit number and then repeat:Divide by 10001: 86528652/10001 = 8652

Select any 2 digit number and then repeat: Divide by 101: 3939/101 = 39

The examples above using inductive reasoning do not prove the hypothesis. I shall prove the hypothesis by using a six digit repitive number tuvtuv. We can represent the number as follows:

$100{,}000t + 1{,}000u + 100v + 100t + 10u + v$ = a six digit number

Add the numbers = $100{,}100t + 10{,}010u + 1001v$

Divide by 1001 and we get: $100t + 10u + v$ = the number that repeats

Chapter 45.0 Solving Linear Simultaneous Equations

Four Methods to solve Simultaneous Linear Equations in Euclidian Space:

Chapter 45.1 Solving Two-Dimensional Linear Simultaneous Equations by Substitution

The first method involves substitution.

$$3X - 2Y = 20$$

$$-5X + 9Y = 12$$

Isolate X or Y $X = \dfrac{12 - 9Y}{-5}$

Substitute in the other equation

$X = 3(12 - 9Y) / (-5 - 2Y = 20)$

Simplify by multiplying both sides by -5.

$3(12 - 9Y)/-5 + 10Y = -100$

$-17Y = -136$, $Y = 8$,

Substitute Y in either equation to find X

$3X - 2(8) = 20$, $3X = 36$, $X = 12$

Chapter 45.2 The second method involves subtraction

$3X - 2Y = 20$
$-5X + 9Y = 12$

Multiply each side by an integer to obtain common coefficients

$5(3X - 2Y) = 5(20)$
$3(-5X + 9Y) = 3(12)$
$15X - 10Y = 100$
$-15X + 27Y = 36$
$17Y = 136$
$Y = 8$

Substitute in first equation

$3X - 2(8) = 20$
$3X = 36$
$X = 12$

Chapter 45.3 The third method involves linear algebra and matrices.

$3X - 2Y = 20$

$5X + 9Y = 12$

Set up two matrices by placing coefficients into upper and lower matrices. The numerical values of the matrices determine the values of X and Y. The bottom matrix is called the "determinant."

The determinant matrix is the same for both X and Y.

$$Y = \frac{\begin{vmatrix} 3 & 20 \\ -5 & 12 \end{vmatrix}}{\begin{vmatrix} 3 & -2 \\ -5 & 9 \end{vmatrix}} \qquad X = \frac{\begin{vmatrix} 20 & -2 \\ 12 & 9 \end{vmatrix}}{\begin{vmatrix} 3 & -2 \\ -5 & 9 \end{vmatrix}}$$

But first we have to learn how to expand a 2 × 2 and a 3 × 3 matrix to obtain a numerical value:

The numerical value of a 2 × 2 matrix, a, b, c, d is calculated as follows:

$$\begin{vmatrix} a & b \\ c & d \end{vmatrix}$$

Value = $(a \times d) - (b \times c)$

$$Y = \frac{(3)(12) - (20)(-5)}{(3)(9) - (-2)(-5)} \qquad X = \frac{(20)(9) - (-2)(12)}{(3)(9) - (-2)(-5)}$$

$$Y = \frac{36 + 100}{27 - 10} \qquad X = \frac{180 + 24}{27 - 10}$$

$$Y = 9 \qquad X = 12$$

Chapter 45.4 Determinant = 0, For Parallel Lines

If both lines are parallel, the determinant = 0. Here is the proof: Shown below are the equations for two lines in standard format. "m" is the slope and "b" is the Y intercept.

The slopes (m). of the two lines are equal.

$$Y - m1X = b1$$
$$Y - m2X = b2$$

The Determinant = $\quad 1 - m_1$
$\qquad\qquad\qquad\quad 1 - m_2$

$D = (1)(-m2) - (-m1)(1)$; if $m_1 = m_2$ then $D = 0$

Chapter 45.5 Solution if slopes are of opposite signs.

Determinant = the sum of the slopes

$$Y - m_1 X = b_1$$
$$Y + m_2 X = b_2$$

Let $m_1 = -m_2$; $D = (1)(m_2) - (-m_1)(1) = m_2 + m_1$

Expanding the matrices, we get

$$Y = \frac{(b_1)(m_2) + (b_2)(m_1)}{m_2 + m_1} \qquad X = \frac{(1)(b_2) + (1)(b_1)}{m_2 + m_1}$$

If the lines are perpendicular then the slopes are negative reciprocals.

Chapter 45.6 Graphical Solution

The fourth method involves plotting the two equations onto an X-Y plane. Each equation describes a ine and the intersection of the two lines determines a singular point which represents the solution for X and Y.

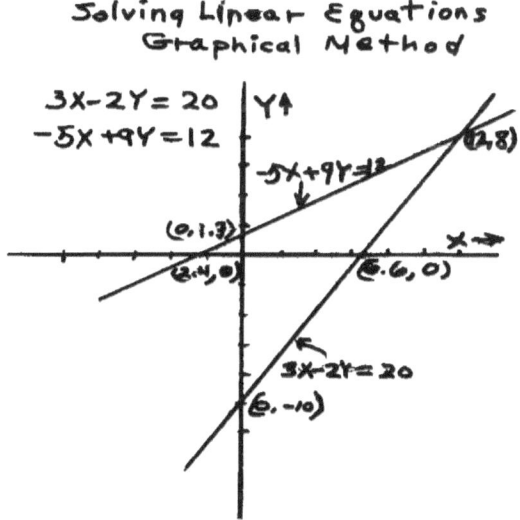

Chapter 45.7 Solving 3 dimensional linear equations

Method to solve three simultaneous linear equations in Three Dimensional Euclidian Space using Linear Algebra and Matrices:

$$3X - 2Y + 5Z = 70$$

$$-5X + 9Y + 7Z = 82$$

$$7X + 8Y - 6Z = 88$$

As before, but this time we set up three matrices as follows: But first we have to learn how to expand a 3 × 3 matrix to obtain a numerical value. Assume a matrix as follows:

$$\begin{vmatrix} a & d & g \\ b & e & h \\ c & f & i \end{vmatrix}$$

Numerical value is as follows:

$a(ei - hf) - b(di - gf) + c(dh - ge) = $ value

$$\frac{\begin{vmatrix} 70 & -2 & 5 \\ 82 & 9 & 7 \\ 88 & 8 & -6 \end{vmatrix}}{\begin{vmatrix} 3 & -2 & 5 \\ -5 & 9 & 7 \\ 7 & 8 & -6 \end{vmatrix}} = X \quad \frac{\begin{vmatrix} 3 & 70 & 5 \\ -5 & 82 & 7 \\ 7 & 88 & -6 \end{vmatrix}}{\begin{vmatrix} 3 & -2 & 5 \\ -5 & 9 & 7 \\ 7 & 8 & -6 \end{vmatrix}} = Y \quad \frac{\begin{vmatrix} 3 & -2 & 70 \\ -5 & 9 & 82 \\ 7 & 8 & 88 \end{vmatrix}}{\begin{vmatrix} 3 & -2 & 5 \\ -5 & 9 & 7 \\ 7 & 8 & -6 \end{vmatrix}} = Z$$

Note that the determinant is the same for each variable. Each matrix has to be expanded into three matrices to solve for X, Y, and Z. Let us first expand the first row of the determinant which equals:

$$3 \begin{vmatrix} 9 & 7 \\ 8 & -6 \end{vmatrix} - 5 \begin{vmatrix} -2 & 5 \\ 8 & -6 \end{vmatrix} + 7 \begin{vmatrix} -2 & 5 \\ 9 & 7 \end{vmatrix}$$

$= 3(-54 - 56) + 5(12 - 40) + 7(-14 - 45)$

$= -330 + (-140) + (7 - 413) = -883 =$ Determinant

Now expand the top matrix for variable X.

$$70 \begin{vmatrix} 9 & 7 \\ 8 & -6 \end{vmatrix} - 82 \begin{vmatrix} -2 & 5 \\ 8 & -6 \end{vmatrix} + 88 \begin{vmatrix} -2 & 5 \\ 9 & 7 \end{vmatrix}$$

$= 70(-54 - 56) - 82(12 - 40) + 88(-14 - 45)$

$= -7700 - (-2286) + (-5192) = -10{,}596$

$X = (-10{,}596)/(-883) = 12$

Now expand the top matrix for variable Y.

$$3 \begin{vmatrix} 82 & 7 \\ 88 & -6 \end{vmatrix} - 5 \begin{vmatrix} 70 & 5 \\ 88 & -6 \end{vmatrix} + 7 \begin{vmatrix} 70 & 5 \\ 82 & 7 \end{vmatrix}$$

$= 3(-492 - 616) - 5(-420 - 440) + 7(490 - 410)$

$= -3324 - 4300 + 560 = -7074$

$Y = (-7064)/(-883) = 8$

Now expand the top matrix for variable Z.

$$3 \begin{vmatrix} 9 & 82 \\ 8 & 88 \end{vmatrix} -5 \begin{vmatrix} -2 & 70 \\ 8 & 88 \end{vmatrix} +7 \begin{vmatrix} -2 & 70 \\ 9 & 82 \end{vmatrix}$$

=3 (792 - 656) +5 (-176 - 560) + 7(-164 - 630)

= 408 - 3680 - 5558 = -8830

Z = (-8830)/(-883) = 10

Therefore the results are X = 12, Y = 8, & Z = 10 The diagram below shows the intersection point.

Resultant $R^2 = X^2 + Y^2 + Z^2$ R = 17.549

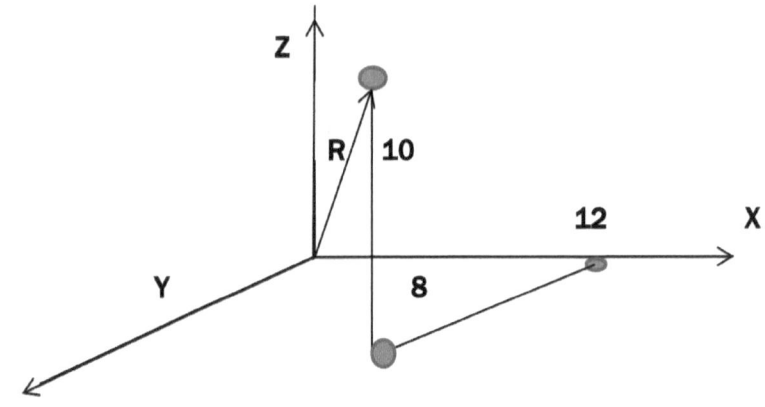

Expanding a 3 × 3 matrix using the first row (and not the first column): Using the same three equations on page

$$\begin{vmatrix} a & d & g \\ b & e & h \\ c & f & i \end{vmatrix}$$

Numerical value is as follows:

a (e I - h f) - d (b I - h c) g (b f - g c) =

$$3 \begin{vmatrix} 9 & 7 \\ 8 & -6 \end{vmatrix} - (-2) \begin{vmatrix} -5 & 7 \\ 7 & -6 \end{vmatrix} + 5 \begin{vmatrix} -5 & 7 \\ 7 & 5 \end{vmatrix}$$

= 3(-54 - 56) + 2 (30 - 49) + 5 (-40 - 63)

= - 330 - 38 - 515 = - 883

The determinant – 883 is the same determinant as the previous example. Values for X, Y, & Z can be expanded around the first row and provide the same answers.

Chapter 45.8 Example of solving an electrical circuit with 3 unknowns

The example below demonstrates the usefulness of solving three simultaneous equations to define a simple electrical circuit. There a few rules you first have to know:

An electrical circuit is defined by values of voltage, electrical current in amperes, resistance of components in ohms (ohms = volts/amperes). Ohms are designated by the Greek letter omega.

Kirchoff's Law states that the sum of electrical currents entering and leaving a node = zero.

Entering current is assigned to be plus and leaving is assigned to be minus.

Kirchoff's Law states that the sum of voltages around a closed loop = zero. The voltage across power sources (batteries and generators) are drawn from minus to plus and the voltage across resistances are drawn from plus to minus.

The simple circuit below shows four resistors and two batteries connected by electrical wires. There are two loops. The three currents are defined as i1, i2, & i3. We have to find the currents and voltages in the two loops. The three equations with three unknowns are solved by using matrices.

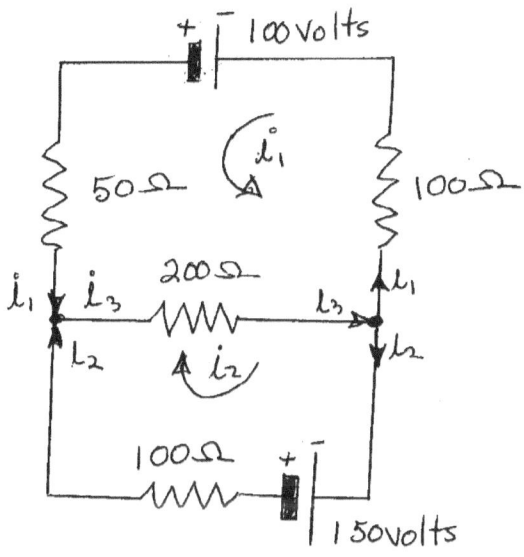

① $i_1 + i_2 = i_3$

② $-100 + 50i_1 + 200i_3 + 100i_1 = 0$

③ $-150 + 100i_2 + 200i_3 = 0$

Re-write

$i_1 + i_2 - i_3 = 0$
$150 i_1 + 0 i_2 + 200 i_3 = 100$
$0 i_1 + 100 i_2 + 200 i_3 = 150$

Three simultaneous Equations

$$\begin{array}{ccc} i_1 & i_2 & i_3 \end{array}$$

$$\begin{vmatrix} 1 & 1 & -1 & 0 \\ 150 & 0 & 200 & 100 \\ 0 & 100 & 200 & 150 \end{vmatrix}$$

Expand first row

$$1\begin{vmatrix} 0 & 200 \\ 100 & 200 \end{vmatrix} - 1\begin{vmatrix} 150 & 200 \\ 0 & 200 \end{vmatrix} - 1\begin{vmatrix} 150 & 0 \\ 0 & 100 \end{vmatrix}$$

$$1(-200)(100) - 1(150)(200) - (150)(100)$$

$$-20,000 - 30,000 - 15,000$$

Determinant, $D = -65,000$

Expand first column

$$1\begin{vmatrix} 0 & 200 \\ 100 & 200 \end{vmatrix} - 150\begin{vmatrix} 1 & -100 \\ 100 & 200 \end{vmatrix} + 0\begin{vmatrix} 1 & -1 \\ 0 & 200 \end{vmatrix}$$

$$-20,000 - 45,000$$

$$D = -65,000$$

Calculating i_1

$$\begin{vmatrix} 0 & 1 & -1 \\ 100 & 0 & 200 \\ 150 & 100 & 200 \end{vmatrix}$$

$$-100 \begin{vmatrix} 1 & -1 \\ 100 & 200 \end{vmatrix} + 150 \begin{vmatrix} 1 & -1 \\ 0 & 200 \end{vmatrix}$$

$$-100(200+100) + 150(200)$$

$$-30,000 \qquad + 30,000 = i_1 = 0$$

Calculating i_2

$$\begin{vmatrix} 1 & 0 & -1 \\ 150 & 100 & 200 \\ 0 & 150 & 200 \end{vmatrix}$$

$$1 \begin{vmatrix} 100 & 200 \\ 150 & 200 \end{vmatrix} - 1 \begin{vmatrix} 150 & 100 \\ 0 & 150 \end{vmatrix}$$

$$-10,000 - 22,500$$

$$-32,500$$

$$i_2 = \frac{-32,500}{-65,000} = 0.50$$

Calculating i_3

$$\begin{vmatrix} 1 & 1 & 0 \\ 150 & 0 & 100 \\ 0 & 100 & 150 \end{vmatrix}$$

$$-10,000 - 1 \begin{vmatrix} 150 & 100 \\ 0 & 150 \end{vmatrix}$$

$-10,000 - 22,500$

$-32,500$

$i_3 = \dfrac{-32,500}{-65,000} = 0.50$

$i_1 = 0.00$
$i_2 = 0.50$
$i_3 = 0.50$

100 Volts

0 volts

0 volts

100V i_3

i_2

50 Volts

150 volts

Chapter 46.0 Dropping a stone through the earth

The acceleration above the earth obey's Newton's Law of Gravity. The force of gravity is inversely proportional to the distance from the center of the earth. The force is governed by the following equation:

$F = (M1\ M2\ G)/R^2$ where m1 is the mass of the earth and m2 is the mass attracted to the earth and G is the universal Gravitational Constant. R is the distance from the center of the earth to the mass.

However, what happens to the force of gravity within the earth. At the center of the earth, the forces of attraction are equal so we expect the force of gravity to be zero. We shall perform a Thought Experiment (Gedanken Experiment) in which we imagine a hole from one side of the earth to the opposite side. There is no air or magma or anything to obstruct a stone falling toward the center. We just drop a stone and are able to record its path toward the center and watch it emerge from the other side.

We have to consider that as the stone is attracted to the center, the earth's mass above the stone is pulling it back and reducing the gravitational pull toward the center. We assume that the density of the earth is constant and the mass = density × volume. The volume of the cap through the stone falls is shown below: The mass above the stone may be considered as a kind of dark matter pulling back the stone just as cosmologists believe that dark matter suffuses the universe and affects the velocities of stars within galaxies. Stars at the outer rim of our galaxy are travelling faster than predicted by Newton's Laws. That is possible if the mass of the galaxy is greater than the measured values.

Dimensional Analysis

M1 = Mass = Newtons/meters/second² or Kg   R = distance in meters

F = force = Newtons   G = 6.67 × 10¹¹ meters³ /(Kg – seconds²)

Therefore F = Kg meter/seconds² = Newtons

From Newton's Laws: Velocity = acceleration × time, acceleration may not be constant

Distance traveled in the direction of the center of the earth = ½ g t² , t = time and g = accel. of gravity

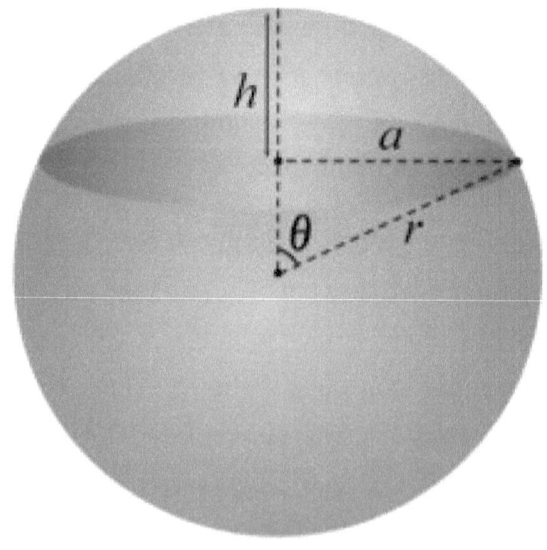

[Abstract]

$V = \pi h^2(3r-h)/3$

V = volume   mass = density times volume

Density is assumed to be constant throughout the Earth.

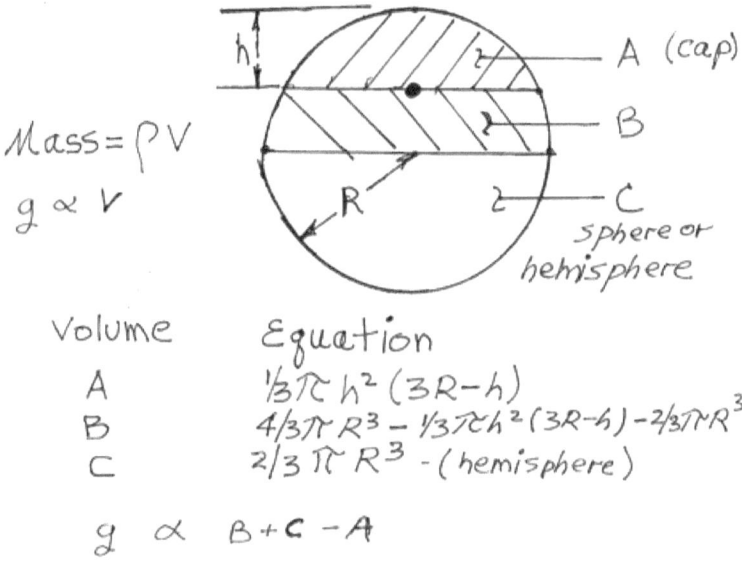

The diagram above shows the cap, mid-section, and hemisphere of the earth. The dark dot is the falling stone. The sections of the earth are divided into A, B, & C. The force of gravity pulling the stone are sections B&C. Section A is The volume of each section is known. The force pulling the stone down = volume of B + C. Volume A pulls the stone back. The force of gravity pulling the stone toward the the acceleration of gravity at the earth's surface = 32.2 ft/sec².

The equation below shows the relationship between the acceleration of gravity within the earth as a function of h/R(depth /radius of the earth). After the stone passes through the center the force pulling the stone forward = C −B − A because B and A are pulling the stone backwards. The stone has enough velocity travelling through the center with zero gravity to emerge at the other end with zero velocity just as it was dropped with zero velocity thus preserving the Law of Conservation of Energy.

g (acceleration of gravity) proportional to $(4/3)\pi R^3[4 - 2(h/R)^2(3 - (h/R)]$

$g = 32.2 [1 - ½ (h/R)^2(3 - h/R)]$

If h = 0, then g = 32.2 ft/sec² If h = R, g = 0

H = R at the center of the earth.

For the return trip past the center

$g = 16.1 [(h/R)^2 (3 - h/R) - 2]$

If h/R = 1 (Center of earth), g = 0.

If h/R = 0, which represents the top of the earth on the other side, then g = -32.2 ft/sec² because the stone is pulled backwards so g is negative.

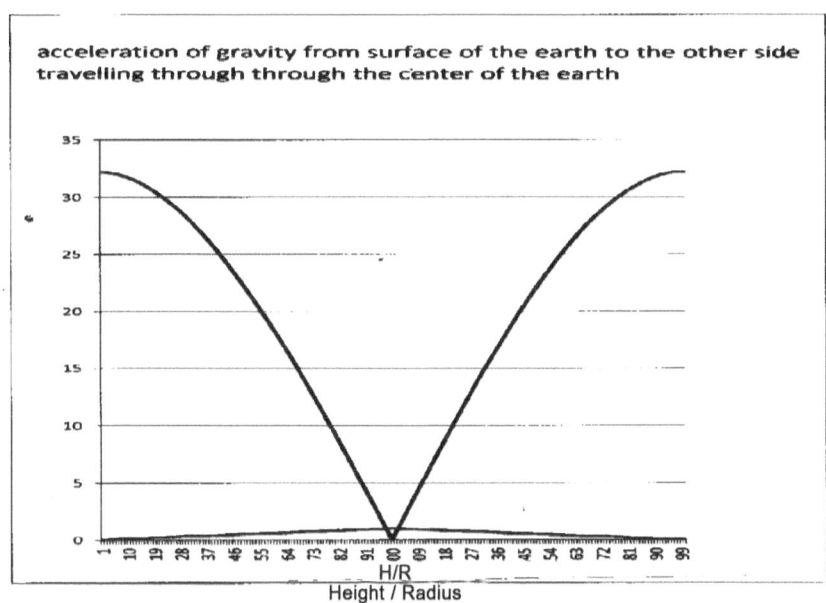

How long does it take to travel to the center of the earth? The velocity = acceleration × time. The acceleration for each increment of h/R has to be calculated. The time to transit the earth from the surface to the center is shown below, wherein g is the acceleration of gravity at each h/R. R = radius of the earth = 3900 miles or 20.59×10^6 feet. h/R goes from zero (surface of the earth) to the center where h/R = 1.

The vertical axis is the time in seconds to reach the center

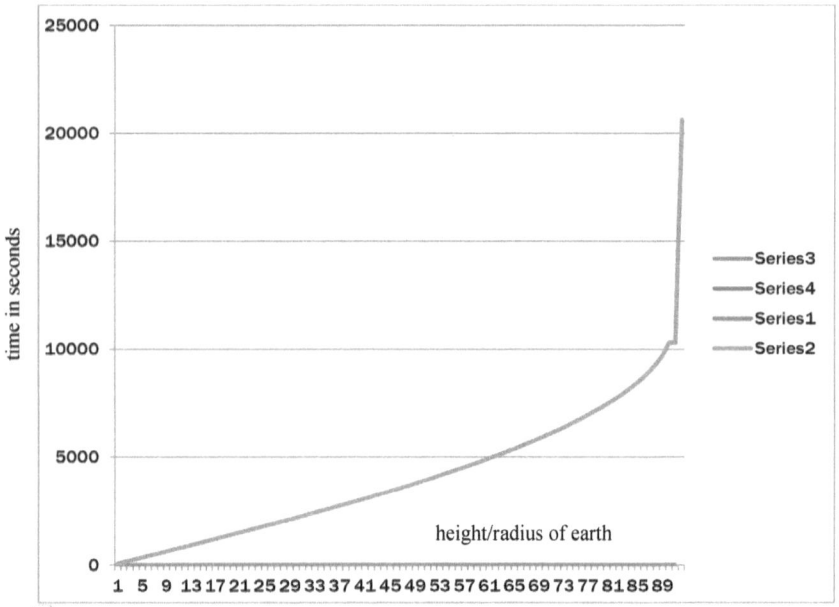

Time to reach the center = 21,000 seconds or 5.8 hours. Velocity slows down as the stone approaches the center. The slope of the line defines the velocity as the stone approaches the center of the earth. Slope = approximately $(0.9-0.61) \times 3900$ miles $\times 5280$ feet/mile $/(10,000-5,000)$ seconds = 1194 feet/sec or 814 miles per hour.

Gravity is zero at the earth's center. The stone has enough velocity to continue to move toward the opposite surface. The force on the stone decreases because the distance between the stone and the earth's center increases. The stone continues until gravity pulls the stone in the opposite direction to where it emerges with zero velocity at the other end of the hole.

Chapter 47.0 Computer Accuracy Check:

$(111,111,111)^2 = 12345678987654321$

If your computer performs this calculation incorrectly, then your computer is defective and untrustworthy. Hand held calculators have 8-10 windows to display numbers. A computer needs 17 windows to display the complete answer. There is a shorter version: $(111,111)^2 = 12345654321$

If your hand-held calculator performs this calculation incorrectly, then your hand-held calculator is defective and untrustworthy.

Chapter 48.0 Quiz

Quiz failed by most college graduates:

A ball and bat cost $110.00.

The bat cost $100.00 more than the ball. How much did the ball cost?

Answer to Quiz Question:

Let X = Cost of ball.

Let Y = Cost of bat.

X + Y = $110.00

X + $100.00 = Y

Substitute for Y

X + X + $100 = $110; 2 X = $10.00

X = $5.00; Y = $105.00

Chapter 49.0 Cryptography Introduction

This section is important because all messaging from your computer transmitted to your bank, Amazon Account, School Account is protected from hacking by Cryptographic methods.

This section provides the students with a rudimentary understand of methods used to protect computer transmissions from eves dropping, access, and false authentication.

The fundamental challenge of cryptography is that Alice wants to send a message to Bob without it being intercepted, read and/or modified by Eve. (Eve is a third person or computer capability)

Corollary 1: Bob wants to read and send a return message to Alice without it being intercepted, read and/or modified by Eve.

Corollary 2: Alice and Bob want assurances that the messages sent and received are really from each of them. Alice and Bob need not be people. Alice can query her bank account using a password and expects that the account recognizes her and no one else. She expects that Eve cannot access the account.

Traditional Assumptions:

1. Alice and Bob are separated in space and/or in time.
2. Alice encrypts the plain text message into a cipher text using a key.
3. Bob decrypts the message using the same key called a symmetric key.

4. Eve has to capture the message and acquire the key in order to break down the cipher text into plain text in reasonable time.

Here is a simple example of cryptography use: We have a safe whose combination is 25 50 25.

We desire to post an encrypted safe combination in public place enabling only selected persons to open the safe without having to memorize a long number. The key to open the safe is an easy number to remember and it is buried in the long posted number.

Only selected people are told the key which 1776.

Security Container Combination:	255025	Called Plain Text Correct combination
Add key	1776	Called Key
Encrypted Combination:	256801	Encrypted Text is posted in a public place

The algorithm consists of adding the key to the plain text combination.:[K + P = E]. The Office Manager permits the posting of encrypted safe combinations. Does this meet the definition of cryptography? Algorithm is simple, quick, and reversible. Key (easy four digit number) is known by few authorized people. The encrypted combination to is therefore: 256801 (public) - 1776 = 255025.

The key is known only by authorized personnel. Authorized personnel simply subtract 1776 from 25680 to acquire the correct combination to open the safe.

Chapter 49.1 Time needed to open a 6 number combination lock

How long will it take to break the three pairs of numbers of encrypted plaintext?

A combinations lock has numbers such as 25 50 25 [six digits]

Each digit or integer can range from 0 to 9. Each integer can be one of 10 integers.

The number of possible numbers is $10 \times 10 \times 10 \times 10 \times 10 \times 10 = 10^6$

If each attempt takes 5 seconds, the maximum time to find the correct combination is 5 seconds/60seconds/min×hour/60mins × days/24 × 10^6 = 58 days. This is the maximum expected time required to test all the combinations. There is a chance however that the correct combination might be determined well before every combination has been tested.

Chapter 49.2 Example:

Using a key for brief message:

Alice wishes to send Bob a message.

Key used by Alice involves substitution of each letter with another letter:

Plain letters: A B C D E F G H I J K L M N O P Q R S T U V W X Y Z

Cipher letters: D E F G H I J K L M N O P Q R S T U V W X Y Z A B C

Alice sends: Plain text: SEE   YOU   MONDAY.

Bob receives: Cipher text: VHH   BRX   PRQGDB.

Called "Caesar Shift Cipher" or homophonic substitution perhaps this is the method used by Julius Caesar to communicate messages to his legions. Each plain text letter is replaced by another letter.

Every plain text letter is advanced by 3 letters with wrap-around.

Algorithm: C = P + 3 (Modulo 26 because there are 26 letters in the English alphabet) is reversible because Bob subtracts the key to read the message correctly. Key = 3.

More about Modulo math later !

These are the disadvantages for Alice and Bob in using a "Caesar Shift Cipher" to encrypt a message.

1. Algorithm & key have to be known by Alice and Bob
2. For long messages, cipher letters are counted and frequency analysis leads to code breaking.
3. Can use unique language rules (e.g. "u" follows "q" and "i" before "e" except after "c", etc) to decode.
4. Alice and Bob have to communicate to select an algorithm, key, and make any changes.
5. Time needed to encrypt plain text and decrypt cypher text.
6. If Eve intercepts long messages, she may use frequency analysis for decryption.

Chapter 49.3 Frequency analysis for English Language:

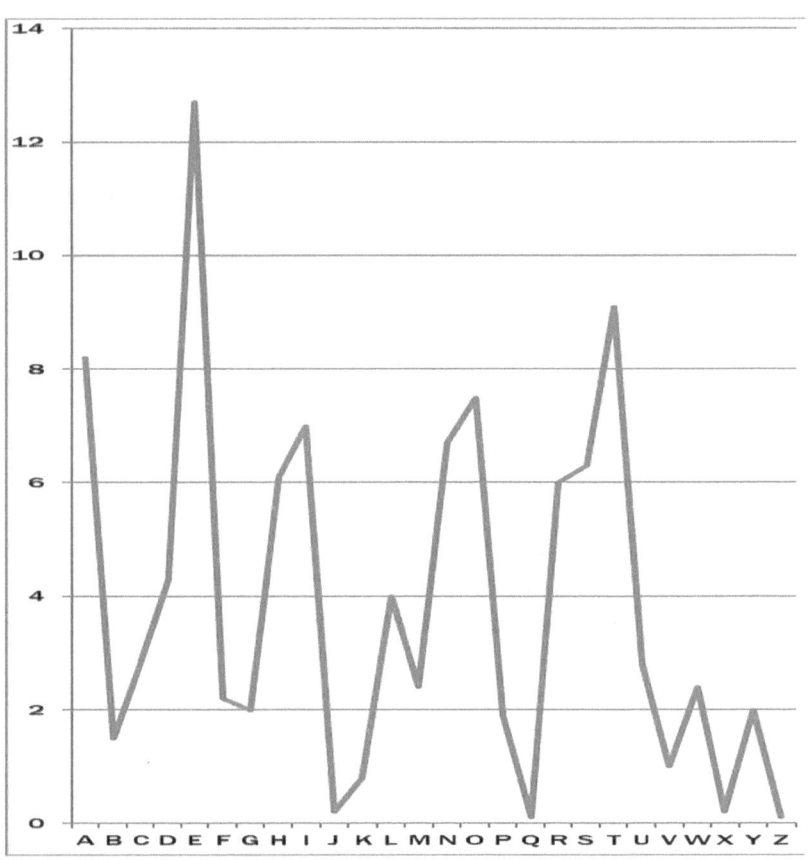

Certain letters can be discounted because their frequency is = or < 1% such as J, K, Q, V, X, & Z.

We need a flat frequency code or cipher to confuse code breakers.

Code breakers use letter frequency to break codes.

letter	percentage
A	8.2
B	1.5
C	2.8
D	4.3
E	13
F	2.2
G	2
H	6.1
I	7
J	0.2
K	0.8
L	4
M	2.4
N	6.7
O	7.5
P	1.9
Q	0.1
R	6
S	6.3
T	9.1
U	2.8
V	1.0
W	2.4
X	0.2
Y	2.0
Z	0.1

The cypher key for any long message can be broken when certain letters appear more often than others. For example the letter "E" appears most of the time in any long message. So if "E" is replaced by "V" and "V" appears most of the time in the message, then code breakers would assume that "V" replaces "E" in the encrypted message.

Chapter 49.4 Leslie Groves Cypher

Here is a flat frequency checkerboard cipher used by General Leslie Groves. He was the Project Director of the Los Alamos Atomic Bomb development during WWII. Each letter of plaintext is replaced by a two digit number: (Partial Polyphonic Substitution) (Reference 7) Several numbers shared the same letter to confuse code breakers. His messages required protection.

	1	2	3	4	5	6	7	8	9	0
1|	I_8	P	I		O	U	O		P	N
2|	W	E	U	T	E	K_6		L	O	
3|	E	U	G	N	B_4	T	N		S	T
4|	T	A	Z_2	M	D		I	O	E	
5|	S_9	V	T	J		E		Y		H
6|	N_7	A	O	L	N	S	U	G	O	E
7|		C	B	A	F	R	S_5		I	R
8|	I	C	W	Y_3	R	U	A	M		N^0
9|	M	V	T		H_0	P	D	I	X	Q
0|	L	S	R_1	E	T	D	E	A	H	E

One-time cipher pads (called ciphers) were changed often and differed for different people. Codes were locked in a safe. Code was not used for entire message; only few letters/word.

Cipher text: 93 31 64 28 07 70 39 02 62 81 97 58 60 66

Plain Text: TELLER SAID YES! Although plain text has repeating letters, cipher text has no repeating numbers.

To further confuse Eve, the numbers would be listed without spaces in groups of five and could be published in an open source with a grouping of numbers that define a recipient.

Example:

67801 93316 42807   67801 defines a recipient.
93316 42807 denotes   78 on the cypher table denotes
TELLER.   no letter, it is a space.

Chapter 49.5 Polyphonic Substitution

Although frequency flat, this method is labor and time intensive. This is an example of polyphonic substitution with numbers. Each letter replaced by different numbers.

Using a book as a key, select a book and use a different page for each day. Here is an example of polyphonic substitution:

key: w e h o l d t h e s e t r u t h s t o b e s e l f e v I d e n t.
       a b c d e f g h i l m n o p r s t u w y

i omitted the letters with frequency = < 1%

    j, k, q, v, x & z

Alice and Bob know the rules and make the appropriate polyphonic substitutions (each letter of the cipher text maps to several same letters of the plain text). b maps to e, i maps to e. This is still labor intensive and subject to successful crypto analysis if the messages are long. Computers are now used for crypto-analyses. Different algorithms or keys are needed to make code breaking difficult and yet easy for Alice to transmit and Bob to receive and decrypt without consuming much time.

Chapter 49.6 Random Letter Substitution

(Use key with or without algorithm)

For 26 different letters, each letter mapped to another letter. (Homophonic substitution)

Example: "A" may be mapped to 26 letters including itself

        "B" may be mapped to 25 letters since A is mapped

        "C" may be mapped to 24 letters since A & B are mapped

Combinations or number of keys = $26 \times 25 \times 24 \ldots = 26! = 4 \times 10^{26}$ (Each letter is used only once in this key)

ADVANTAGES FOR ALICE AND BOB

1. Many combinations are available.

2. Can change combination daily if necessary.
3. Time to decode all combinations at 0.001 microseconds per combination = age of universe.

DISADVANTAGES FOR ALICE AND BOB

1. Need at least 2 code books showing letter substitutions (one each for Alice and Bob)
2. Multiple code books needed to send multiple simultaneous messages - risk of capture of code books
3. Need prior agreement when to meet and change codes
4. Use of language rules assists code-breakers
5. Frequency analysis may be used to assist code-breakers
6. Time is needed to encrypt and to decrypt

Chapter 49.7 Thomas Jefferson's Cypher Wheel

Thomas Jefferson is considered the Father of American Cryptography. He invented a wheel cipher or code system (note spelling) using a cylinder with 36 wheels on a central axis. Each wheel was inscribed with 26 letters and 10 numbers=36 alphanumeric characters. The alphanumeric characters on each wheel were jumbled. They were not alphabetical.

The operator rotates each wheel to find the appropriate letter and construct a message

along one line not exceeding 36 alphanumeric characters. Alice transmits a line of jumbled letters to Bob. Bob uses his identical wheel cypher to lay out the identical jumbled letters and then reads the plain text.

Number of combinations = 36! = 371,993,326,789,901,217,467,999,448,150,835,200,000,000

The Jefferson disk, or wheel cypher as Thomas Jefferson named it, also known as the Bazeries Cylinder, is a cipher system using a set of wheels or disks, each with the 26 letters of the alphabet arranged around their edge. The order of the letters is different for each disk and is usually scrambled in some random way. Each disk is marked with a unique number. A hole in the center of the disks allows them to be stacked on an axle. The disks are removable and can be mounted on the axle in any order desired. The order of the disks is the cipher key, and both sender and receiver must arrange the disks in the same predefined order. Jefferson's device had 36 disks. [Kahn, p. 194]

Once the disks have been placed on the axle in the agreed order, the sender rotates each disk up and Down until a desired message is spelled out in one row. Then the sender can copy down any row of text on the disks other than the one that contains the plaintext message. The recipient simply has to arrange the disks in the agreed-upon order, rotate the disks so they spell out the encrypted message on one row, and then look around the rows until he sees the plain text message, i.e. the row that's not complete gibberish. There is an extremely small chance that there would be two readable messages, but that can be checked quickly by the person coding.

First invented by Thomas Jefferson in 1795, this cipher did not become well known and was independently invented by Commandant Etienne Bazeries, the conqueror of the Great Cipher, a century later. The system was used by the United States Army from 1923 until 1942 as the M-94. This system is not considered secure against modern code breaking if it is used to encrypt more than one row of text with the same ordering of disks (i.e. using the same key). *See#* Cryptanalysis.

Even if the cipher were computerized, and each combination Even if the cipher were computerized and each combination completed in one picosecond (10^{-12} second) the time required to complete all combinations = 3.71×10^{29} seconds. Jefferson never pushed it. I suspect he found it cumbersome to use. Jefferson's cipher text is labor intensive to encode and decode.

But there is another issue. How does a computer recognize plain text? First, the computer has to identify a key and then identify the number of letters in a word. It uses a computer dictionary with perhaps 1,025,109 entries (including variants because of tense, prefixes, plurals, etc.) and compares the cipher text with the entries in a dictionary. First we have to confuse the code breaker by making a key difficult to decrypt. We can substitute one letter for another. How many possible ways are there to permute two letters, three letters, four letters etc…up to 10 letters? The number of ways to permute 26 letters is as follows:

N = 26! = 26 × 25 × 24 etc. = 26! = 4.0329×10^{26}. Using a 34 petaflop (peta floating point calculations per second) = 34×10^{15} would require $4.0329 \times 10^{26}/34 \times 10^{15} = 1.1186 \times 10^{10}$ seconds.

Seconds per year = 31,536,000 or 300 years. Each plain text letter has a different cypher letter assigned to it.

Chapter 50.0 Combination and Permutations Reminder:

Code makers are interested in creating a key with the maximum number of combinations to thwart anyone trying to break the code and establish the key. The more combinations and permutations of letters makes it more difficult for code breakers.

Assume an alphabet of four letters: A, B, C, & D

How many ways can we combine them taking two at a time? If AB, AC, AD, BC, BD and CD are Interchangeable (i.e. AB = BA, AC = CA, etc., then there are six possibilities. (on your hand calculator enter 4 nCr 6) This means that from 4 letters we can make 6 two letter words where order is important.

Some of the words may not be in the dictionary.

If AB not = AB, & AC not = CA etc., then there are 12 two letter words where order is not important.

(On your hand calculator enter 4nPr2).Some of the words may not be in the dictionary. Most two-letter words are not in the dictionary.

Our English alphabet consists of 26 letters.

How many possible two letter words, where order is not important, can we obtain from our 26 letter alphabet? How many three, four, five, six, seven, and eight letter words can we make from our 26?

word alphabet? Code breakers need to know this because they want to know where there are spaces between words. We have 12 possible words but not all are in the dictionary. Most two-letter possible words are not in the dictionary.

Number of possible 2, 3, 4, 5, 6, 7, 8, & 9 letter words:

	26 nPr	= Result
Possible words with one letter	26 nPr 1	= 26
Possible words with two letters:	26 nPr 2	= 650
Possible words with three letters	26 nPr 3	= 15,600
Possible words with four letters:	26 nPr 4	= 358,800
Possible words with five letters:	26 nPr 5	= 7,893,600
Possible words with six letters:	26 nPr 6	= 165,765,600
Possible words with seven letters:	26 nPr 7	= 3,315,312,000
Possible words with eight letters:	26 nPr 8	= 62,990,928,000
Possible words with nine letters:	26 nPr 9	= 1,133,367,048,000

The total number or words having 1 to 9 letters = 1.2×10^{12} words.

The number of words in an English dictionary is approximately 171,500, which is a small percentage of all the possible words. There is ample possibility to replace real words with letters.

Chapter 50.1 Word Counts

Distillation of the Google books data gives us 97,565 distinct words, which were mentioned 743,842,922,321 times. To no surprise, the most common word is "the". Here are the top 50 words, and their overall percentage, This distribution gives a code breaker an idea of words expected to be found in a long encrypted message.

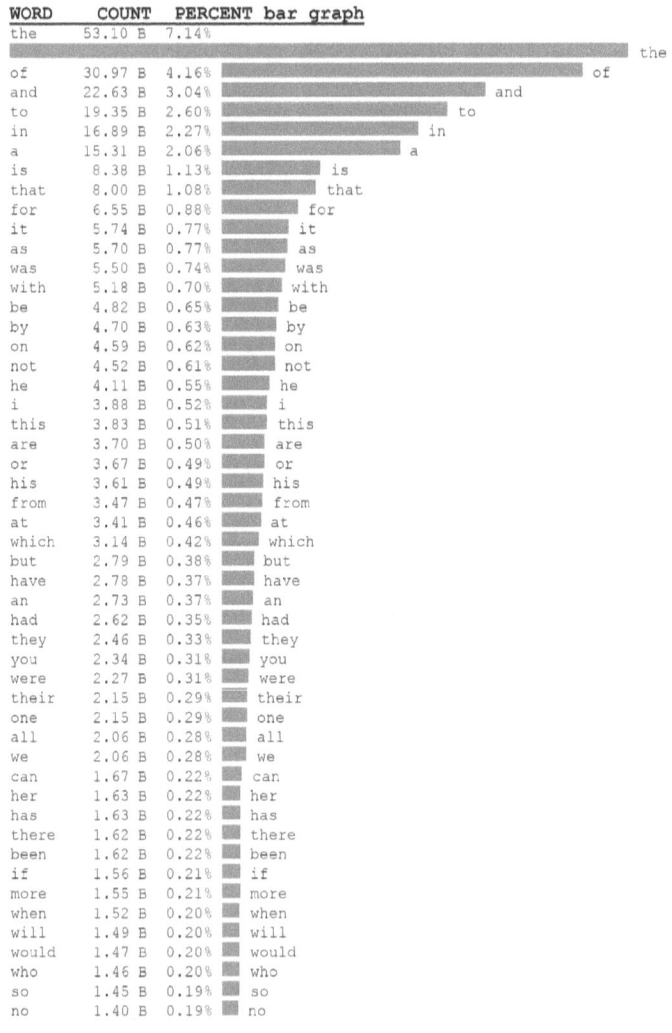

WORD	COUNT	PERCENT
the	53.10 B	7.14%
of	30.97 B	4.16%
and	22.63 B	3.04%
to	19.35 B	2.60%
in	16.89 B	2.27%
a	15.31 B	2.06%
is	8.38 B	1.13%
that	8.00 B	1.08%
for	6.55 B	0.88%
it	5.74 B	0.77%
as	5.70 B	0.77%
was	5.50 B	0.74%
with	5.18 B	0.70%
be	4.82 B	0.65%
by	4.70 B	0.63%
on	4.59 B	0.62%
not	4.52 B	0.61%
he	4.11 B	0.55%
i	3.88 B	0.52%
this	3.83 B	0.51%
are	3.70 B	0.50%
or	3.67 B	0.49%
his	3.61 B	0.49%
from	3.47 B	0.47%
at	3.41 B	0.46%
which	3.14 B	0.42%
but	2.79 B	0.38%
have	2.78 B	0.37%
an	2.73 B	0.37%
had	2.62 B	0.35%
they	2.46 B	0.33%
you	2.34 B	0.31%
were	2.27 B	0.31%
their	2.15 B	0.29%
one	2.15 B	0.29%
all	2.06 B	0.28%
we	2.06 B	0.28%
can	1.67 B	0.22%
her	1.63 B	0.22%
has	1.63 B	0.22%
there	1.62 B	0.22%
been	1.62 B	0.22%
if	1.56 B	0.21%
more	1.55 B	0.21%
when	1.52 B	0.20%
will	1.49 B	0.20%
would	1.47 B	0.20%
who	1.46 B	0.20%
so	1.45 B	0.19%
no	1.40 B	0.19%

Chapter 50.2 Letter Counts

Enough of words; let's look at letter counts. There were 3,563,505,777,820 letters mentioned. Here they are in frequency order:

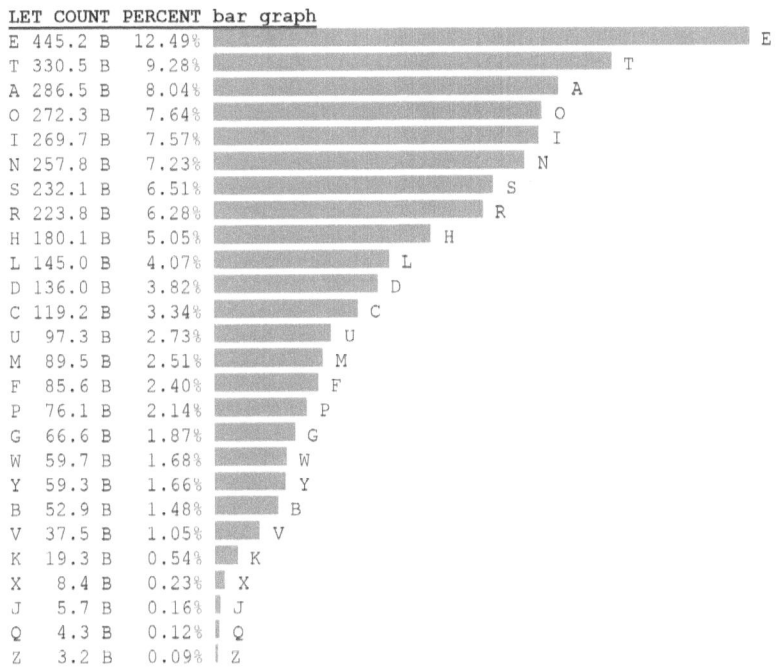

```
LET COUNT PERCENT bar graph
E 445.2 B  12.49%                                            E
T 330.5 B   9.28%                                  T
A 286.5 B   8.04%                                A
O 272.3 B   7.64%                               O
I 269.7 B   7.57%                               I
N 257.8 B   7.23%                              N
S 232.1 B   6.51%                            S
R 223.8 B   6.28%                           R
H 180.1 B   5.05%                       H
L 145.0 B   4.07%                    L
D 136.0 B   3.82%                   D
C 119.2 B   3.34%                 C
U  97.3 B   2.73%              U
M  89.5 B   2.51%             M
F  85.6 B   2.40%             F
P  76.1 B   2.14%            P
G  66.6 B   1.87%           G
W  59.7 B   1.68%          W
Y  59.3 B   1.66%          Y
B  52.9 B   1.48%         B
V  37.5 B   1.05%       V
K  19.3 B   0.54%    K
X   8.4 B   0.23%  X
J   5.7 B   0.16%  J
Q   4.3 B   0.12%  Q
Z   3.2 B   0.09%  Z
```

Here is the distribution for distinct words (that is, counting each word only once regardless of how many times it is mentioned). Now the average is 7.60 letters long, and 80% are between 4 and 10 letters long:

Code breakers know to look for words between 5-8 letters long.

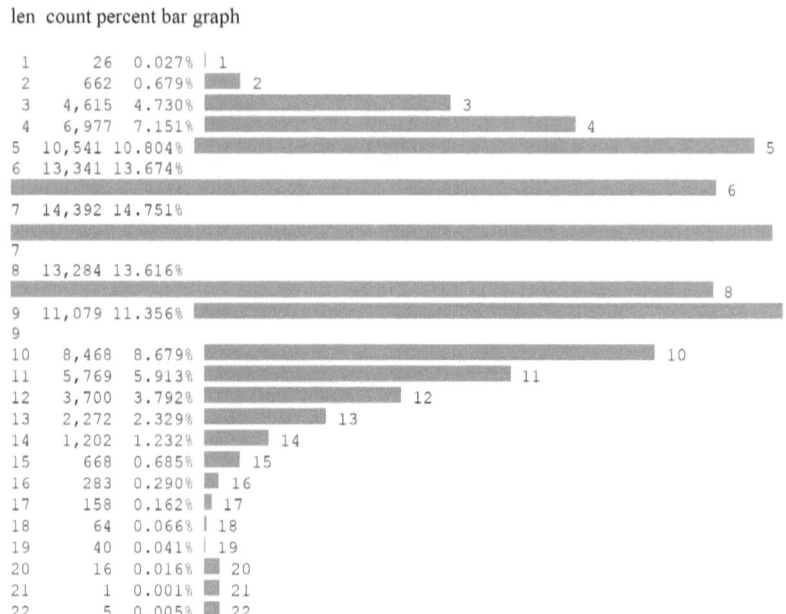

```
len  count  percent  bar graph
 1      26   0.027%
 2     662   0.679%
 3   4,615   4.730%
 4   6,977   7.151%
 5  10,541  10.804%
 6  13,341  13.674%
 7  14,392  14.751%
 8  13,284  13.616%
 9  11,079  11.356%
10   8,468   8.679%
11   5,769   5.913%
12   3,700   3.792%
13   2,272   2.329%
14   1,202   1.232%
15     668   0.685%
16     283   0.290%
17     158   0.162%
18      64   0.066%
19      40   0.041%
20      16   0.016%
21       1   0.001%
22       5   0.005%
23       2   0.002%
```

Different algorithm or key needed to confound code-breakers and speed process of decryption which is still labor intensive. The computer has a word dictionary with about 750,000 words.

How to decode letters grouped as follows:

VHHBR XPRQG DBURY DSYUT RULMN STREQ XIZPQ

NBDER REWQU OPUTMNLOMB UTLKM APLMQ PUYSE

Average length of a word = 5 letters. Decoder selects 5 letters and fast computer attempts to make words listed in computer dictionary. Use grammar rules. Subject comes first followed by verb and predicate.

Chapter 50.3 Random Number Homophonic Substitution

Replace letters in plain text with 3 digit random numbers such as: A = 014, B = 456, etc. The number of combinations of 26 numbers (one per letter) from a total of 999 numbers is:

$999! = 1.7 \times 10^{51} = 2^{170.2}$ Called 170 bit -more later! 26! (999 - 26)!

This method uses a key to assign random numbers for each letter. There are 1.7×10^{51} possible keys.

Advantages for Alice and Bob

1. Can change combination daily if necessary
2. Time needed for Eve to try all combinations
3. There are many possible keys and at 0.001 microseconds per combination, the time required for crypto-analysis exceeds age of universe.

Disadvantages for Alice and Bob

1. Need at least 2 codebooks showing number substitutions. Multiple codebooks needed to send multiple messages - risk of capture of code books
2. Need prior agreement when to change codes Frequency analysis may still be used for numbers that appear more frequently
3. Random number generator required
4. Labor and time intensive

Estimate that a 100 MHz INTEL PENTIUM with 8 MB RAM requires 50 years to factor

10^{130} or $2^{431.85}$

Chapter 60.0 Binary Numbers

Binary Numbers were invented by Gottfried Wilhelm Leibnitz in 1703 as a mathematical curiosity. Binary numbers now form the basis of all communication devices and computers. Each character on the keyboard is defined by a binary number consisting of a string of zeros and ones. Binary numbers are uses to encrypt and decrypt messages. Binary numbers are of base 2.

Any integer can be composed of powers of base 2; e.g.

19	$= 1 \times 2^4 +$	0×2^3	$+$	0×2^2	$+$	1×2^1	$+$	1×2^0
	$= 16$	$+ 0$		$+ 0$		$+ 2$		$+ 1$
	$= 1$	$+ 0$		$+ 0$		$+ 1$		$+ 1$
19	$= 10011$							

Base 10	Binary (Base 2)	Powers of base 2	
0	0	0	
1	1	2^0	
2	10	2^1	
3	11	$2^1 +$	2^0
4	100	2^2	
5	101	$2^2 +$	2^0
6	110	$2^2 + 2^1$	

7	111	$2^2 + 2^1 + 2^0$
8	1000	2^3
9	1001	$2^3 + 2^0$
10	1010	
11	1011	
12	1100	
13	1101	
14	1110	
15	1111	
16	10000	2^4
17	10001	
18	10010	
19	10011	
20	10100	
21	10101	
22	10110	
23	10111	
24	11000	
25	11001	
26	11010	
27	11011	
28	11100	
29	11101	
30	11110	
31	11111	$2^4 + 2^3 + 2^2 + 2^1 + 2^0$
32	100000	2^5

A stream of binary digits may represent any number in base 10. The binary digit "1" may be represented by a pulse and the binary digit

"0" may be represented by an absent pulse. This representation may be inverted in which "0" is represented by a pulse and "1" by an absent pulse. Other engineering representations include positive and negative changes in frequency from an established mean frequency to represent "0" and "1." Numbers increase in size from right to left.

Chapter 60.1 ASMAD for binary numbers

Addition of Binary Numbers

$$12 = 1100$$
$$+\ 13 = 1101$$
$$25 = 11001 \quad 1 + 1 = 10 \text{ carrying } 1$$

Multiplying Binary numbers:

```
  12      1100
 ×13     ×1101
 156      1100
         0000
         1100
         1100
      10011100 = 156
```

Subtracting Binary numbers:

0 – 1 =   - 1 (have to borrow)

```
 101       1100101
-75       -1001011
 26        0011010
```

Division by Binary Numbers

There are no decimals so we have to multiply numerator by a factor of ten or more, and then divide the final answer:

Example: Compute 16/5. Multiply 16 by 1000. Compute 16000/5 and divide answer by 1000.

5 = 101 in Binary

16000 = 11111010000000 in Binary

$$= \frac{110010000000}{101}$$

```
      ┌─────────────────
 101  │ 11111010000000
        101
        ───
         101
         101
         ───
           101
           101
           ───
           0000000
```

UNTAUGHT MATH   113

Answer = 110010000000 = 3200 Then divide by 1000 = 3.2 So we get 16/5 = 3.2

Chapter 60.2 ASCII

ASCII stands for American Standard Code for Information Interchange. Computers can only understand numbers, so an ASCII code is the numerical representation of numbers, letters, and/ or characters.

Hex means hexadecimal and are numbers with base 16. Capital "A" = 65 and in binary, it is 1000001 = 64 + 1 = 65. Each BIT represents a keyboard character. Capital B = 66.

Upper Case Letters

Description	Char	Hex	Character
Commercial "at" sign	**@**	**40**	**@ → @ $1000000 = 2^6$**
capital A	A	41	A → A $1000001 = 2^6 + 0^5 + 0^4 + 0^3 + 0^2 + 0^1 + 2^0$
capital B	B	42	B → B $1000010 = 2^6 + 0^5 + 0^4 + 0^3 + 0^2 + 2^1 + 0^0$
capital C	C	43	C → C 1000011
capital D	D	44	D → D 1000100
Capital E	E	45	E → E 1000101
capital F	F	46	F → F 1000110
capital G	G	47	G → G 1000111
capital H	H	48	H → H 1001000

capital I	I	49	I → I 1001001
capital J	J	4A	J → J 1001010
capital K	K	4B	K → K 1001011
capital L	L	4C	L → L 1001100
capital M	M	4D	M → M 1001101
capital N	N	4E	N → N 1001110
capital O	O	4F	O → O 1001111
capital P	P	50	P → P 1010000
capital Q	Q	51	Q → Q 1010001
capital R	R	52	R → R 1010010 = $2^6 + 0^5 + 2^4 + 0^3 + 0^2 + 2^1 + 0^0$
capital S	S	53	S → S 1010011
capital T	T	54	T → T 1010100
capital U	U	55	U → U 1010101
capital V	V	56	V → V 1010110
capital W	W	57	W → W 1010111
capital X	X	58	X → X 1011000
capital Y	Y	59	Y → Y 1011001
capital Z	Z	60	Z → Z 1011010

Each ASCII capital letter has 7 bits or binary digits.

One byte = 8 bits. A nibble = 4 bits. A computer word may consist of 8, 16, or 32 bits. Usually there are 7 bits per computer word. The 8th bit, if used, is a parity bit meaning that if the number of ones are even, then the parity bit may be zero. If the number of ones are odd, then the 8th bit may be one. This is done to detect errors in transmission of bits A keyboard has 128 separate characters. 2^7 = 128 bits which are enough to capture all symbols on a standard keyboard.

Chapter 60.3 Reversible Algorithms

So what is the problem? The problem has shifted from breaking algorithms and keys to distribution of keys and the long duration needed to encode and decode.

New algorithms are needed to quickly transmit keys over the internet and still confound code breakers.

Each letter (except the rarely used letters) is assigned a number with six bits (10^6) combinations.

Alice sends same message to Bob using ASCII but adds a key or selected binary number) to each transmitted ASCII binary number. Bob knows the key and subtracts the key from each group of five or more binary numbers. Alice and Bob have worked this out in advance.

Binary digits may be added, subtracted, multiplied, and divided to confound code breakers.

The algorithms have to be reversible meaning that if Alice adds binary numbers to the ASCII message,

Then Bob has to be able to subtract the binary numbers added by Alice. The binary numbers added or subtracted from the ASCII message is the encryption or decryption key. key.

Add key to plain text and subtract key from cipher text

Multiply plain text by key to and divide cipher text by key

Use of really difficult equations: Simple example follows:

C = cipher text, P = plain text, K = key = 8

We want to transmit the letter "S".

P = Plain Text = "Letter S" = 83 = 1010011 [ASCII assigned number]

C = (P + k) (P + 2 k) = (83+8) (83+16) = 9009 = 10001100101001

Bob has to decrypt by solving quadratic equation for "P"

P = 0.5 {-3 k + [(3k)^2 - 4 (2 k^2 − C)]$^{1/2}$} = 83 = 1010011 = "S"

But only quadratic, cubic, and some quartic equations can be solved in closed form. This one is not so easily solved:

C = (P + k) (P + 2k) (P + 3 k) (P+4k) (P+5k) etc...

NSA works to create polynomials that are not easily solved but which their mathematicians have solved. Cryptographers want irreversible algorithms. Crypto-analysts want reversible algorithms. There is a tension between protection of transmitted messages and ease of decryption of a transmitted message.

Chapter 60.4 Reversible and Irreversible algorithms

Alice wants to send an encrypted message over the internet to Bob and wants Bob to be able to easily decrypt the message knowing that Eve may have intercepted the encrypted message.

Bob needs a key and an ASMAD algorithm to decrypt the message. Protection is maximized if the ASMAD algorithm and the key are transmitted separately in person.

Alice has two ways to encrypt messages to Bob.

1. Assign each of the 26 letters of the alphabet to a number between 1 and 999. There are 7.0×10^{77} combinations. [Press 999 nPr 26 on your hand-calculator]. Alice has software that modifies each letter of the Plane Text and converts it into cypher text. Bob had encryption software that allows him to decrypt the cypher message into plane text. The software algorithm is reversible.
2. Use ASCII. Each letter or symbol is assigned a number. There are about 128 characters in ASCII and one character is a parity bit so an ASCII word has 8 bits. Modify all the 8 bits of each number with a reversible algorithm and a key. Use ASMAD as an algorithm: Here are some examples:

Add 00110010 (= 50) to every ASCII bit in the plane text. Add is the algorithm and 50 is the key.

Subtract 00110010 (= 50) from every ASCII bit in the plane text. Subtraction is the algorithm and 50 is the key.

Multiply each bit of the plane text by 50. Multiplication is the algorithm and 50 is the key.

Divide each bit of the plane text by a key. Have to inform Bob about the remainder and divisor for each divided bit. This step is more complex but is more difficult to decrypt because two keys are needed, the remainder and the divisor.

Chapter 60.5 Swapping BITS

Use of large numbers with a simple algorithm can delay decryption. Encrypting a plain text message in ASCII

Alice wants to send Bob "SEE YOU MONDAY."

S	E	E	Y	O	U
1010011	1000101	1000101	1011001	1001111	1010101

M	O	N	D	A	Y
1001101	1001111	1001110	1000100	1000001	1011001

Swap 2nd and 3rd bits: BITS are counted from right to left.

1010101	1000110	1000011	1011001	1001111	1010011
1001011	1001111	1001110	1000010	1000001	1011001

Not every BIT was changed because some 2nd and 3rd BITS were 1 1 or 0 0.

Other methods are as follows:

1. Square each plain text binary number.
2. Combinations of squaring and adding binary numbers.
3. Add numbers from letters of another document such as Preamble of US Constitution.
4. Add ASCII of each letter of "We the People…" to plain text message.
5. Swap bits in plain text and/or cypher text; e.g. swap 2nd and 4 bits.

6. Multiply each plain text number by a large number such as a googol.

$OO = \text{Googol} = 10^{100} = 2^{332.1928094}$

Advantages for Alice and Bob

1. Alice and Bob memorize simple algorithm; key code book not necessary if one key is used
2. Frequency analysis more difficult
3. Can use both letter substitutions and algorithms but need key for letter substitutions.
4. Use of fast computers available to encrypt plain text and decrypt cypher text since we are using bits

Disadvantages for Alice and Bob

1. Multiple code books needed to send multiple messages-risk of capture of code books
2. Changes in key have to be communicated
3. Have to separate the stream of bits or create algorithm that separates them – bits have different lengths.

Chapter 60.6 IBM Encryption Technology

1. IBM developed "Lucifer"- sophisticated cipher algorithm
2. Long binary strings broken into blocks of 64 bits
   Bits of each block are shuffled and rearranged
3. Details of shuffling determined by key
4. Lucifer beyond code breaking capability of NSA

5. NSA lobbied Congress to limit key to 56 bits.
6. 56 bit key version adopted Nov 75 and called Data Encryption Standard or DES –Today DES may be broken within 24 hours.

Key has 2^{56} combinations = $2^{10} \times 2^{10} \times 2^{10} \times 2^{10} \times 2^{10} \times 2^{6}$

= (1024) (1024) (1024) (1024) (1024) (64)

= (1073741824) (67108864) = approx. $7.205759404+ \times 10^{16}$

DES uses a 64 bit key with 8 bits reserved for parity, hence 56 bits.

DES was replaced by AES (Advanced Encryption Standard) = 128 bits.

Disadvantages for Alice and Bob of using keys

Unavoidable distribution issues

1. Alice and Bob have to meet and exchange keys
2. Lots of couriers needed for multiple hard copy keys

   USG keys managed by NSA for US embassies
3. Eve eavesdrops on lines or on radio waves and intercepts bit stream
4. Need to change and create keys constantly

Disadvantages for Alice and Bob of using algorithms

1. Need to change algorithms or parameters
2. Eve captures cipher-text bits and uses reverse algorithms with fast computers to derive plain text

3. Almost every algorithm is reversible – two way functions even if we try trillions of keys

   Addition vs. subtraction

   Doubling vs. halving

   Squaring vs. square roots
4. Algorithms can be reversed in reasonable time

For more secure encryption we need to avoid sharing keys, we need unbreakable key distribution systems. The use of one-way algorithms (irreversible) can still be decrypted with enough time.

So what is the problem? The problems have shifted from breaking algorithms and keys to distribution of keys and the long time to encode and decode. Challenge appears unsolvable!

Chapter 60.7 Diffie-Hellman Amazing Discovery

Exchanging of keys may not be necessary for security

Example:

Alice sends message to Bob in a small box with padlock.

Bob receives message and adds another lock and resends to Alice.

Alice removes her lock and resends box to Bob.

Bob opens box with his own key.

Alice and Bob do not exchange keys.

EXAMPLE:

Plain text: SEEYOU   MONDAY.

Cipher text: VHH   BRX   PRQGDB.

Bob receives message and uses his own cipher key to encrypt before sending to Alice.

Bob's key: Plain text: A, B, C, D, E, F, G, H, I, J, K, L, etc.
           Cypher text: G, H, I, J, K, L, M, N, O, P, Q, R, etc.

Bob's encrypted message: BNN HXD VXWMJH

Alice decrypts Bob's message with her key YKK EUA SUTJGE

Then she resends message to Bob.

Bob decrypts Alice' message with his key:

"SEE YOU MONDAY"

This discovery forced cryptographers to examine their assumptions.

Advantages for Alice and Bob

1. No exchange or distribution of keys.
2. Each person uses own algorithm and own key or common algorithm and own key.

Disadvantages for Bob and Alice:

1. Eve can still capture message from Alice or Bob and decode.
2. Torturous paths back and forth.

3. Time to send and receive messages twice as long.
4. Algorithms are still reversible.
5. Code-breakers may use frequency analysis and fast computers.

Chapter 61.0 Modulo Math and Irreversible Algorithms

Mathematicians discovered that Modular Arithmetic, (called clock arithmetic) is a mathematical curiosity with interesting properties. Called Modulo (means measure) or Mod.

Consider a clock with 12 numbers.

6 + 3 = 9 (Mod 12)

11 + 4 = 3 (Mod 12) since we go around the clock. There is no 15.

Consider a clock with 7 numbers (Mod 7): 0,1,2,3,4,5,6

6 + 1 = 0 (Mod 7) $3^6 = 1$ **(Mod 7)**

6 + 5 = 4 (Mod 7) $3^7 = 3$ **(Mod 7)**

2 × 6 = 5 (Mod 7) $3^8 = 2$ **(Mod 7)**

$3^2 = 2$ **(Mod 7)** $3^9 = 6$ **(Mod 7)**

$3^3 = 6$ **(Mod 7)** $3^{10} = 4$ **(Mod 7)**

$3^4 = 4$ **(Mod 7)** $3^{11} = 5$ **(Mod 7)**

$3^5 = 5$ **(Mod 7)** $3^{12} = 1$ **(Mod 7)**

Example:

$3^5 = 243$; 243 (Mod 7) = 243/7 = 34 + <u>Remainder of 5</u> = 5.

Output is erratic. Functions are not reversible. Knowing the Mod and the result does not produce the exponent. Could the exponent be used as a means of specifying a key?

Mathematicians Whitfield Diffie, Martin Hellman, and Ralph Merkle at Stanford University discovered a key exchange method allowing Alice and Bob to exchange key information without Eve being able to use what she intercepts to compute the key. System works as follows:

Phase	Alice	Bob	Eve
1.	Selects a Secret Number 3 = X	Selects a Secret Number 6 = Y	Hears nothing!
2	Selects Mod = 11	Selects same Mod = 11	Overhears Mod = 11
3	Selects base = 7	Selects same base = 7	Overhears base = 7
4	Selects exp. function 7^X (Mod 11) 7^3 (Mod 11)* = 2	Selects same exp. function 7^Y (Mod 11) 7^6 (Mod 11) = 4	Overhears exp. functions Unaware of X & Y Unaware of 3 & 6
5.	Tells Bob Answer = 2	Tells Alice Answer = 4	Overhears 2 & 4
6.	Computes	Computes	Does nothing

	4^3 (Mod 11)	2^6 (Mod 11)	Does nothing
	= 9	= 9	Can't compute
7.	key = 9	key = 9	Can't compute key

Eve can't work backwards to compute key!

8. This is the key used to alter the plain text.

*Example: 7^3 (Mod 11) = 343/11 = 31 + Remainder of 2

Summary of what happened

Eve intercepted everything except the two secret numbers selected by Alice and Bob.

Eve had the answers to the modulo calculations but could not reverse the process to compute the secret numbers and compute the key.

In reality the numbers used are much larger than the numbers in this example. Excel limits us to ten digit displays. The computed key would contain many more digits.

Alice and Bob still had to agree upon and communicate the common modulo number and the common Base number for an exponential function although they do not have to privately meet to exchange keys.

It does not matter if Eve overhears the modulo number and common base number because the irreversible algorithm does not allow Eve to compute the key.

The same key is used to encrypt the plain text and to decrypt the cypher text. Key is called a symmetric key.

A remarkable breakthrough!

Diffie, Hellman, and Merkle looked for an asymmetric key system unique to Alice and unique to Bob that would provide absolute protection against decoding by Eve and not require communication to exchange information between Alice and Bob. Was this possible?

What Diffie, Hellman, and Merkle wanted:

1. Alice creates and publishes an irreversible public key that anyone may use to encrypt messages to her. The public key should be irreversible so that no one can decrypt messages sent to Alice including the sender.

Chapter 61.1 RSA Algorithm Public & private key cryptography

April 1977, enter three mathematicians at MIT Lab for Computer Science:

Ronald Rivest, Adi Shamir and Leonard Adelmen invented a one-way asymmetric cipher (called the RSA asymmetric cipher) that satisfies all criteria. Eventually, RSA Data Security Corporation in Bedford, Massachusetts was incorporated to commercialize the RSA cipher.

The RSA process and algorithm developed into product called Pretty Good Privacy. "User is provided with two separate keys, one widely distributed public key encrypts the information, while the private key, held only by one person, is the only key that unlocks the encryption algorithm."

A disadvantage of the RSA asymmetric encryption algorithm was the large computing power and time needed for encryption and decryption. A symmetric encryption scheme requires less computing power and less time.

Outline of RSA algorithm. Now we can understand how it works and remove the mystery.

Alice wants to send Bob the message "See You Monday." She needs to know Bob's public key.

On her computer, RSA is loaded, she types the message to the recipient (Bob) and then presses the command to "encrypt." Prior to this event, Bob has loaded RSA on his computer. Here is what happens: (I use four small prime numbers for easy computation.)

Alice	Bob	Eve (hacker)
Selects 2 large prime numbers secretly: a, b	Selects 2 large prime numbers secretly: p, q	Detects nothing
use 19 & 29	use 17 & 11	
Multiplies prime numbers	Multiplies prime numbers	Does nothing
(19)(29) = 551	(17)(11) = 187	

Releases large number sends to Bob 551	Releases large number sends to Alice 187	Learns large numbers 551 & 187
Selects exponent = 5	Selects exponent = 7	Learns exponents 5 & 7
Public keys Product = 551 exponent = 5	Public keys Product = 187 exponent = 7	Provided with Public keys
Private key primes: 19, 29	Private key primes: 11, 17	Does not know private primes

Alice wants to send a plain text message to Bob

"SEE YOU MONDAY", "S" is the first letter to be encrypted.

Both have to compute their private decryption keys, s & t, to exchange information.

From the math developed by RSA; the values of s and t are different for Alice and Bob.

Alice	Bob	Eve
Encryption key (exponent) times (s) = 1 {Mod (a-1) (b-1)} (19-1)((29-1) = 504	Encryption key (exponent) times (t) = 1 {Mod (p-1)(q-1)} (17-1)(11-1) = 160	Can not compute keys.

UNTAUGHT MATH

$5s = 1 \{\text{Mod } 504\}$ $7t = 1 \{\text{Mod } 160\}$ Equations are secret and not shared

Since $504/5 = 100.8$ Since $160/7 = 22.85$
$S = 101$ $t = 23$ Decryption keys are unknown.

Private decrypt key   Private decrypt key

Standard ASCII Codes

0	00	00000000	NUL	null
1	01	00000001	SOH	start of header
2	02	00000010	STX	start of text
3	03	00000011	ETX	end of text
4	04	00000100	EOT	end of transmission

Alice transmits to Bob   Eve may intercept

Using Bob's Public keys   but cannot read message

Sends: $C = 83^7 \{\text{Mod } 187\}$

$C = 8$ using private key

From Alice to Bob

So Alice sends Binary number $= 8$

SCH STX 00001000
ETX EOT
for the first letter
in the message

"SEE YOU
MONDAY."

$P = 8^{23}$ {Mod 187}

Bob decrypts the
message using
his private key

8^t {Mod (p)(q)}

$(17)(11) = 187$

$P = 83$
In ASCII, 83 = "S"
$P = 83 =$ "S" in
ASCII = 00101011
$P =$ Plain text

Eve intercepts
message
SCH STX
00001000
ETX EOT
The binary
code above

00001000 in
ASCII means
backspace.

Bob decrypts the message as follows: SCH STX 00101011 ETX EOT = 83 in ASCII = "S"

It works!!

Chapter 61.2 Calculations using Modulo Math-Table I

Calculation of 83^7 {Mod 187}; $7 = 1 + 2 + 3 + 1$

$83^2 = 157$ {Mod 187}

$83^3 = 128$ {Mod 187}

$83^1 = 83$ {187}

$(8^3 \times 157 \times 128 \times 83)$ {Mod 187} = $(740,328.0428 - 740,328)$ $(187) = 8.000047 = 8$

Calculation of 8^{23} {Mod 187}; $23 = 5 + 5 + 5 + 5 + 3$

$8^5 = 43$ {Mod 187} $8^3 = 138$ {Mod 187}

$(43 \times 43 \times 43 \times 43 \times 138)$ {Mod 187}

$= (2,522,965.44385 - 2,522,965)$ $(187) = 82.99995 = 83$

Chapter 70.0 Prime Numbers Needed for Cryptography

How large a product is needed to discourage code-breaking?

We are talking about the product of two prime numbers. The product of two primes needs to be at least 10^{308} for important banking transactions (The Code Book). Mathematicians are working to factor the products of large primes. In 1977, a product of two primes having 129 digits presented as a challenge. Approx. (10^{129}), was factored in 1994. Can two people have the same prime number?

PGP provides for selection of keys having 512, 2048, or 4096 bits. There are 10^{151} prime numbers from which to choose whose length is less than or equal to 512 bits. The probability of two people having the same prime number, selected from a random process, is less than the probability of all the molecules in a room suddenly moving to one corner depriving the occupants of oxygen.

Chapter 70.1 Privacy and National Security

What about privacy vs. national security?

The intent of Congress Omnibus Crime Bill of 1991 was to ensure the government could obtain plain Text messages of voice, data, etc. when authorized by law to conduct court-authorized investigations. However, commercial, criminal and civil liberty groups lobbied against the measures in digitized Communications since they may be optimized for surveillance by the government for legal means and used by the Criminal element for illegal means.

Chapter 70.2 Origins of PGP (Pretty Good Privacy)

Enter Phil Zimmerman, a private software guru.

His objectives were to modify the RSA algorithm and create a product.

1. To increase speed of the RSA encryption (several minutes required).
2. To increase speed of finding two primes by wiggling mouse randomly.
3. To provide authentication of Alice by means of a digital signature.
4. To receive a license from RSA before launching PGP.

PGP uses new concepts based on the RSA algorithm:

	ALICE	BOB	RESULT
RSA Method	Encrypts message with Bob's public key	Decrypts message with his private key	Privacy
IDEA Method	1st encrypts message with her private key	1st decrypts with his private key and uses 2 sets of keys	Privacy Authentication
	2nd re-encrypts with Bob's public key	2nd re-decrypts with Alice's public key	asymmetric keys
PGP IDEA KEY	1st encrypts message IDEA with her IDEA symmetric key	1st decrypts IDEA key with private key	Privacy and Authentication
	2nd encrypts IDEA key with Bob's public key	2nd decrypts message with IDEA symmetric key	Uses 1 private key and 1 symmetric key

IDEA = International Data Encryption Algorithm uses a 128 bit key.

US Patent no. 5,214,703 expires 5/25/10. Estimated 10^{12} years required to try all keys.

Chapter 70.3 Politics of Protection

In 1993 Phil Zimmerman released PGP to a friend who downloaded PGP onto the internet. Zimmerman was charged with exporting a "munitions system" that permitted terrorists to evade the authority of the US Government.

MIT Press published PGP in a 600-page book. In 1996, the US AG dropped the case. Zimmerman settled with RSA and received a license. In 1997, PGP sold to Network Associates and became a product. PGP is still free to users from the internet but not for commercial use. Web site: http://www.pgpi.com/

Statement by William Crowell, DEPDIRNSA, "If all PCs in the world, approximately 260 million, were to be put to work on a single PGP encrypted message, it would take an average of 12 million times the age of the Universe to break a single message." Use of Quantum Computing may prove this statement wrong because of the expected increase in processing speed.

Remaining Problems and Some Answers:

1. Alice has files of encrypted data in which the government has a legal interest. Can she be forced to release her pass phrase or decrypt her files?

    No nation has a law requiring suspects to decrypt their files or provide a pass phrase except the UK.

2. Alice has access to valuable files and has been captured and coerced into releasing her private key. She may have

disappeared. Foul play is suspected and the law enforcement authorities need to investigate. Her employer is concerned.

Alice could have shared her key with one or more governments or private agencies that could put it together if she was indisposed. Civil libertarians, believers in privacy and strong encryption, did not appreciate this key escrow policy. This policy was appreciated by agencies so the government. There are companies called "Trusted Third Parties" that retain copies of all private keys. Can they be forced to reveal keys after receiving a court order? What happens if a TTP employee breaks confidence and reveals the private key? Can an employer require that an employee share a private key with a TTP?

For many users a TTP would require a huge database for storage of a large number of agreed-on keys.

3. Alice had her identity stolen and Bob is uncertain the message is from Alice. How can he be certain he is receiving her messages?

"Certification Authorities" such as Verisign will verify that a public key corresponds to a particular person. They also verify the validity of digital signatures. Will Alice and Bob agree to post their public keys with a certification authority?

4. Alice is part of a consortium in which multiple approvals from several persons are required before final action is authorized. Can authentication be distributed among several members?

Yes, MITRE is performing research on Secure Distributed Computing. The idea is to distribute private keys or pieces of private keys among several servers, all of which are needed to approve an action that requires authentication by multiple persons.

5. Why is the USG doing Quantum Cryptography (QC)? Development of quantum computing will enable Eve to break public key cryptography because of the speed by which the keys and algorithms can be decrypted. QC is a counter measure to quantum computing since keys can be readily changed for each transmission from Alice to Bob and long transmissions can employ several different keys.

How much time is needed to compromise an encryption method?

Bits	current technology	quantum computing estimate
32	seconds	nanoseconds
56 (DES)	hours	nanoseconds
64	days	milliseconds
128	weeks	milliseconds
256	several weeks	seconds
512	RSA was cracked in 6 weeks	minutes
37,760	Universe will end first	Universe still to end first

Issues have shifted from decoding keys and algorithms to rapid key distribution and authentication.

The Asier™ encryption methods and systems are scalable. Currently, there are six classes of Asier™.

Azier's Encryption algorithm are frequency flat, employ 5,000 to 50,000 bits and produce output to 0s & 1s that are indistinguishable from noise. RSA and Azier algorithms may be used to protect transactions and storage of bits and by tes used for utilities controls, banking and financial transactions, unclassified consumer emails, government classified communications over copper wire, optical cable, or over the air. Encryption may be installed on servers, routers, telephones, radios, notebook computers, pagers, or smart cards. So what is the threat? Breaking the RSA algorithms depend upon factoring large numbers. In Ref 8, the authors estimate that a number of 400 digits could be factored in one year.

Chapter 71.0 Quantum Cryptography (QC)

QC is a method using photons to distribute and regenerate secret keys between remote locations and guaranteed to detect eavesdropping. QC uses properties of light to transmit key and encrypt bits.

Quantum cryptography was invented by Charles Bennet and Giles Brassard in 1984.

Light is transmitted as a stream of separate energy packages called photons. A single photon is called a quantum of light. What are the properties of photons?

1. A stream of photons (collimated) behaves as a ray of light; each photon may be absorbed and/or reflected by an electron.
2. Single photons may not be divided.
3. A single photon may oscillate in many directions (unpolarized) with a single frequency of oscillation.
4. Energy of a single photon=(h)(frequency of oscillation[cps])

h = Planck's constant = 6.5×10^{-27} ergs/cps.

5. Photons may be deliberately forced to oscillate in a preferred direction to by passing them through a crystal such as calcite. This is called polarization.
6. Collimated photons may be filtered (attenuated) to allow one at a time to strike a detector.
7. Photons travel at the speed of light but can be slowed down.
8. Photons avoid each other
9. Collimated photons may not be in phase. Photons emerging from a laser are in phase but may be unpolarized.
10. Any attempt to measure any conjugate property of one photon will disturb some other conjugate property (Heisenberg's Principle).
11. Conjugate properties: circular, rectilinear, or diagonal polarizations

The concept behind QC is for Alice to isolate individual photons and polarize the photons into one of four orientations before transmitting the photons into Bob's receiver. Alice and Bob agree to select zeros and ones to represent each of the four orientations.

- horizontal polarization; call that "0"
- vertical polarization; call that "1"
- plus 45 degrees polarization (diagonal); call that "0"
- Minus 45 degrees polarization (diagonal); call that "1"

Representations

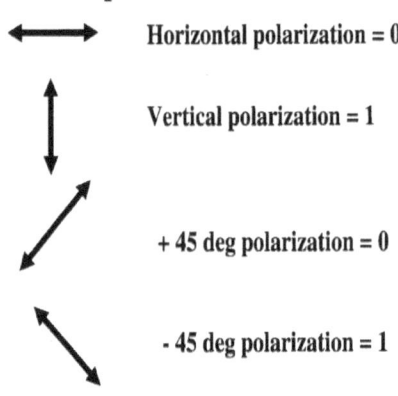

Horizontal polarization = 0

Vertical polarization = 1

+ 45 deg polarization = 0

- 45 deg polarization = 1

Selections are arbitrary and agreed to by Alice and Bob.

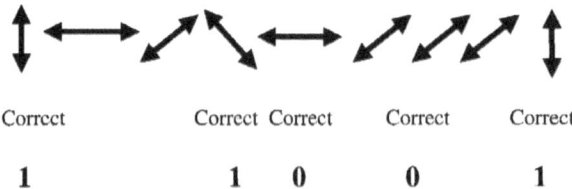

Correct		Correct	Correct		Correct		Correct
1		1	0		0		1

Each polarization is a "0" or a "1" per prior agreement between Alice and Bob. Incorrect bits are discarded.

This is the shared symmetric key between Alice and Bob; 11001. Process appears impossible to crack.

Quantum Symmetric Key Transmission from Alice to Bob

Alice secretly transmits a random sequence of photons in one of four polarizations: (polarizations remain stable in transit). V, H, -45 deg, or + 45 deg.

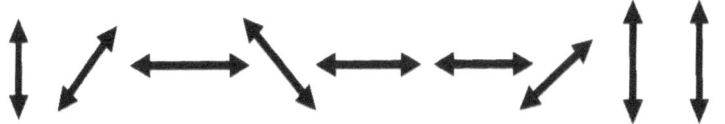

Bob measures the transmission from Alice, using crystals and photo-multiplier tubes to isolate and measure the polarizations. A pair of polarizations is called a basis for the measurement. Bob can only use one basis for each photon; either rectilinear or diagonal. Both basis are randomly used. Some photons may not be received at all.

Alice transmits randomly in secret.

Bob randomly sets his sensor at rectilinear or diagonal basis and tells Alice which basis he used for each photon; either diagonal or rectilinear. He cannot use both basis on the same photon.

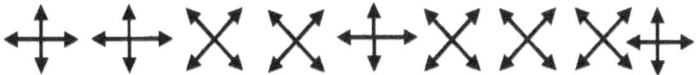

These are Bob's measurements. Alice tells Bob which basis is correct and discards incorrect data.

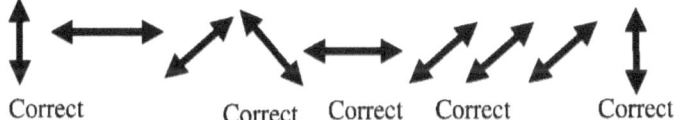

Correct   Correct Correct Correct   Correct

Bob measures the transmission from Alice using polarizing crystals and photo-multiplier tubes to Isolate and measure the polarizations. A pair of polarizations is called a basis for the measurement.

Bob can only use one basis for each photon that is either rectilinear or diagonal. Both bases are randomly used. Some photons may not be received at all.

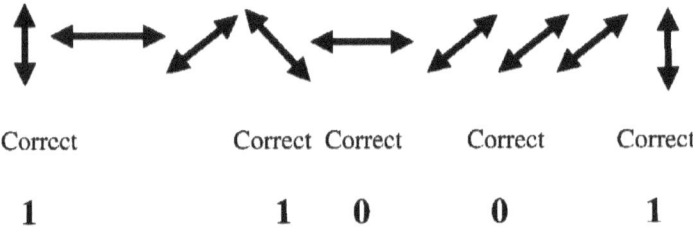

Correct	Correct	Correct	Correct	Correct
1	**1**	**0**	**0**	**1**

Each polarization is a "0" or a "1" per prior agreement between Alice and Bob. Incorrect bits are discarded.

This is the shared symmetric key between Alice and Bob; 11001. Process appears impossible to crack.

:Alice and Bob agree on a key without ever meeting in person.

Eve can only disturb the signal in a way that alerts Alice and Bob. Eve may know the algorithm but not the key.

Quantum encryption does not require use of large prime numbers and/or memory to store the numbers for public and private keys.

Disadvantages:

Is there a risk of compromise?

Eve can intercept the transmission if the pulse contains multiple photons?

Eve cannot measure rectilinear and diagonal polarizations on the same photon without disturbing the photon and causing error bits. However, if the pulses contain multiple photons, Eve could obtain enough photons to make simultaneous rectilinear and diagonal measurements thereby knowing unambiguously the polarizations transmitted by Alice. If Eve knows the polarization scheme, then she gains complete information about the transmission.

Countermeasure:

If Eve captures photons to perform measurements, then fewer reach Bob. Bob would sense the reduction in signal and become suspicious.

Eve can disrupt the transmission if the pulse contains multiple photons.

Eve can capture photons and change the polarizations and then re-insert them into the transmission line.

Bob does not have a clue. The integrity of the system has been compromised since Alice can no longer rely on what she transmitted to Bob.

Countermeasure:

Excessive error bits can be tracked by publicly transmitting a subset of real or dummy data in blocks and comparing "parity" i.e. the evenness or oddness of the bits in the blocks. Disagreements in parity will cause both parties to restart the process.

ID Quantique (Swiss firm) introduced QC products

Chapter 71.2 MagiQ Technologies

MagiQ Technologies is selling QC products including single photon generators and receivers. Optical cables installed under ocean and in use. MagiQ is designing repeaters for underwater use. Building in Virginia optically wired for quantum crypto by MagiQ. Los Alamos doing experiments to send single photons in air Researchers have sent photons over 100 KM optical cable without degradation.

> "NEW YORK, NY – Nov. 3, 2003 – MagiQ Technologies, Inc., *the* quantum information processing (QIP) company, today announced the general availability of its Navajo Secure Gateway, the world's first commercially available quantum key distribution (QKD) system. Relying on the laws of physics rather than the computational difficulty of breaking keys, and easily integrated into existing digital computing infrastructures, Navajo solves key distribution problems that have

been the bane of cryptographers for centuries. Incorporating real-time key generation with quantum distribution of those keys makes for the most secure cryptographic system ever. Navajo offers cost-effective protection from both internal threats, such as disgruntled employees, and external threats including corporate, government, and other sources of exposure. Navajo supports secure key exchange at distances up to 120 km, a major technical accomplishment that makes very long secure spans possible *via* cascading devices."

Chapter 71.3 Recent Developments in Quantum Cryptography

DARPA QUANTUM NETWORK – WORLD'S FIRST QC NETWORK – Reported 23 Oct 2003

Under DARPA sponsorship with Harvard University, Boston University, BBN technologies built and began to operate the world's first Quantum Key Distribution (QKD) network over commercial optical cable. BBN has created high speed detectors and crypto systems based on entangled photons. BBN has identified hacking vulnerabilities and integrated safeguards into the QC Network design. The network runs 24 hours per day. In the near future the universities plan to connect the campuses with QC links to encrypt and decrypt messages. As anyone who works will fiber-based QKD can attest, detectors are the most critical technology at present. Although BBN built its own set of cooled InGaAs APDs for our first two years of development, we have been very fortunate to switch to superb InGaAs detector systems built by Dr. Don Bethune and Dr. William Risk at IBM Almaden for our fiber-based systems.

NEW SPEED RECORD FOR QUANTUM KEY DISTRIBUTION – Reported 18 April 2006

National Institute of Standards (NIST) physicists were able to send single photons at 4 million bits/sec transmitted over 1 km of optical fiber with a bit error rate of 3.6 %. The previous record was 1 million bits/sec transmitted over 1 km. The transmission was successful but slower over 4 km of optical fiber. Phenomena contributing to degradation of the photons are (1) loss of orientation as thee photons move through a curved fiber and (2) spread of the wave function as the photons move through the fiber.

Dead time limits qc speeds– Reported 28 September 2007

Dead time is the period during which the photon detector needs to recover after it detects a photon.

Commercially available single-photon detectors need about 50-100 nanoseconds to recover before they can detect another photon.

New Distance Record for Quantum Key Distribution - Reported 21 July 2009

Researchers from the University of Geneva and Corning Inc. in New York have demonstrated a Quantum Key Distribution (QKD) proto type that can distribute quantum keys over 250 KM in the lab, improving the previous record of 200 km. The researchers used a superconducting photon detector and ultra loss optical fibers made by corning. However, at the longer distances there is a loss of bit rate. At 100 km, the bit rate was 6,000 bits/sec and at 250 km the bit rate was 15 bits/sec with low error rates.

PROTOCOL SHARED REFERENCE FRAMES NOT REQUIRED – Reported 9 Mar 2010

An important requirement of quantum key distribution is that receiver and transmitter have the same reference frame. Linear polarizations will change if transmitter or receiver is in a rotating reference frame. Researchers from the University of Bristol and the national University of Singapore have developed a means of transmission that is independent of their reference frames. Possibilities exist to have transmitter between a moving object in an accelerating reference frame such as an orbiting satellite and a fixed or even mobile terrestrial station.

Chapter 71.4 Quantum Entanglement

Quantum entanglement is a quantum mechanical phenomenon in which the quantum states of two or more objects have to be described with reference to each other, even though the individual objects may be spatially separated. This leads to correlations between observable physical properties of the systems.

For example, it is possible to prepare two particles in a single quantum state such that when one is observed to be spin-up, the other one will always be observed to be spin-down and vice versa, this despite the fact that it is impossible to predict, according to quantum mechanics, which set of measurements will be observed.

Quantum entanglement has applications in the emerging technologies of quantum computing and Quantum Cryptography, and has been used to realize quantum teleportation experimentally. Physicists have succeeded in entangling five photons for the first time. Although four photons have been entangled before, five is the minimum number needed for universal error correction in quantum computation. Entanglement allows particles to have a much closer relationship than is possible in classical physics. For example, two photons can be entangled such that if one is horizontally polarized, the other is always vertically polarized, and vice versa, no matter how far apart they are. Complete information about the quantum state of a particle e is instantaneously transferred by the sender, who is usually called Alice, to a receiver called Bob.

Chapter 72.0 Geometry Rhombus Smallest Perimeter for a parallelogram

Proof that a rhombus has the smallest perimeter for a given area for any parallelogram with a fixed area.

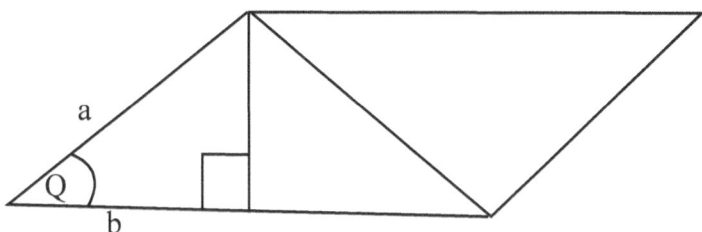

Q is the angle between sides a and b.

Perimeter = P = 2 a + 2 b   h/a = Sin Q

Area of the parallelogram = A = 2 [1/2 b h]

A = {(P - 2 a)/2} {a Sin Q} = {P/2} {a Sin Q} – {2a /2}{a Sin Q}

A = Sin Q {Pa/2 - a²}   This is a parabola.

Take the first derivative of A with respect to "a" to find P minimum and set the derivative = 0.

dA/da = Sin Q{P/2 - 2a} = 0

Therefore P = 4 a. , From Perimeter Equation 4 a = 2a + 2b.

or a = b. This is a rhombus.

Given a parallelogram with Area A, a rhombus of Area A has the smallest perimeter

Chapter 72.1 Smallest perimeter for a Rhombus

Proof that a square has the smallest perimeter for a given area for any rhombus with a fixed area. This is a rhombus.

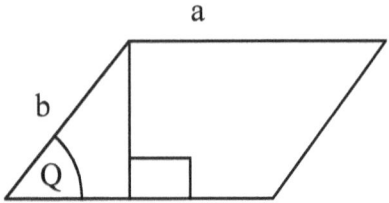

Q is the angle in the corner.

Perimeter = P = 4 a.

Area = A = 2 (1/2 a (a Sin Q)) = {P^2/ 16} {Sine Q}

The area "A" is given. P is a minimum only if Sine Q is a maximum.

Therefore Q = 90 deg. Because Sin of 90 deg = 1.0.

Therefore the rhombus must be a square for P to be a minimum.

Chapter 72.2 Ambiguous Case for Proving Triangles Congruent

Triangles may be proved congruent by the following corresponding parameters if they are equal.

Side   Angle   Side   SAS

Angle   Side   Angl   ASA

Side   Side   Side   SSS

Side Side Angle SSA is called the Ambiguous Case because triangles may not be congruent. See diagram below:

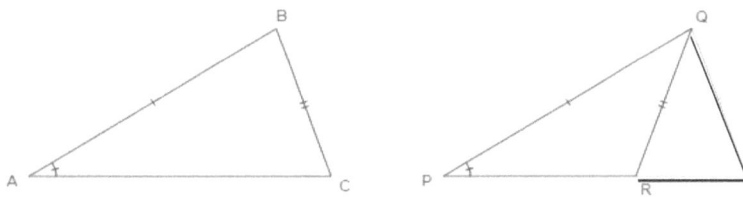

Congruent Triangles - Why SSA doesn't work

PB = PQ, BC = QR Angle P = Angle P This is SSA for both triangles.

Given two sides and non-included angle (SSA) is not enough to prove congruence.

You may be tempted to think that given two sides and a non-included angle is enough to prove congruence. But there are two triangles possible that have the same values, so SSA is not sufficient to prove congruence.

In the figure above, the two triangles above are initially congruent. But if you click on "Show other triangle" you will see that there is another triangle that is not congruent but that still satisfies the SSA condition. AB is the same length as PQ, BC is the same length as QR, and the angle A is the same measure as P. And yet the triangles are clearly not congruent - they have a different shape and size.

Chapter 72.3 Laws of Sine's and Cosines for any triangle

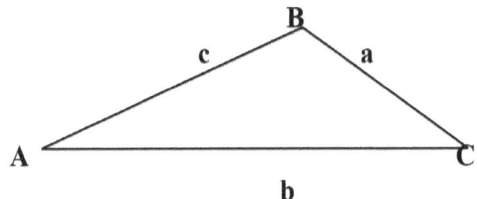

Law of Sine's

$(\sin A) / a = (\sin B) / b = (\sin C) / c$

Chapter 72.4 Finding the 3rd Side of a Triangle Using the (Law of Cosines)

$b^2 = a^2 + c^2 - 2\,a\,c\,\cos A$

$a^2 = b^2 + c^2 - 2\,b\,c\,\cos C$

$c^2 = a^2 + b^2 - 2\,a\,b\,\cos B$

Chapter 72.5 Perfect Right Triangles (called triplets)

$A^2 + B^2 = C^2$ where A, B, & C are Natural Numbers.

Odd no. A	B	C		Even no. A	B	C
3	4	5		4	3	5
5	12	13		6	8	10
7	24	25		8	15	17
9	40	41		10	24	26
11	60	61		12	35	37
13	84	85		14	48	50
15	112	113		16	63	65
17	144	145		18	80	82
19	180	181		20	99	101
21	220	221		22	120	122
23	264	265		24	143	145
25	312	313		26	168	170
27	364	365		28	195	197
29	420	421		30	224	226
31	480	481		32	255	257
33	544	545		34	288	290
35	612	613		36	323	325

Chapter 73.0 Global Positioning System (GPS)

Most automobile drivers use some sort of GPS system in their cars for direction finding.

Most drivers have no idea how it works and the fundamental principles of physics that enable the GPS to provide accurate information about directions and traffic avoidance.

The GPS System uses the principles of Newtonian Physics, General Relativity and Special Relativity.

There are 21 GPS Satellites in orbit 12,427 miles above the surface of the earth.

The global positioning system or GPS is a constellation of satellites in geosynchronous orbit that are used to approximate the location of a GPS receiver on Earth. GPS uses 4 satellites that are in view of the receiver to solve 4 equations for the (x ,y ,z) coordinates of the receiver and a value d which is the difference in time between the receiver's clock and the satellites' clocks. The difference in time is important because the location is approximated using the travel time of a signal from the satellite to the receiver and this happens in much less than a second. The difference in time can cause error because the satellites' clocks are accurate to $10-810-8$ of a second while the receiver›s clock is much less accurate. The equations that are solved to approximate a receiver's location using GPS are:

$$(x-A1)^2 + (y-B1)^2 + (z-C1)^2 - (c(t1-d))^2 = 0$$

$(x-A2)^2 + (y-B2)^2 + (z-C2)^2 - (c(t2-d))^2 = 0$

$(x-A3)^2 + (y-B3)^2 + (z-C3)^2 - (c(t3-d))^2 = 0$

$(x-A4)^2 + (y-B4)^2 + (z-C4)^2 - (c(t4-d))^2 = 0$

x, y, and z are the rectangular coordinates of the receiver, A, B, and C. A, B, and C are the Coordinates of the satellites, d is the difference in time between the receiver and the satellites clocks and it is the travel time for the signal from the satellite to the receiver. The approximate location of a receiver is obtained by solving the system of equations for specific satellite coordinates using the quadratic equations. The four equations have to be solved simultaneously.

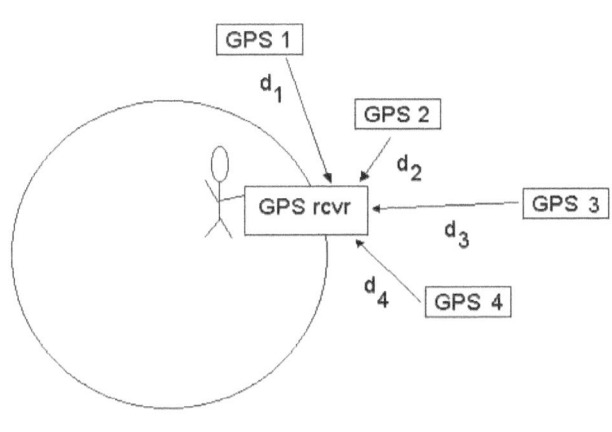

(x−A2)2+(

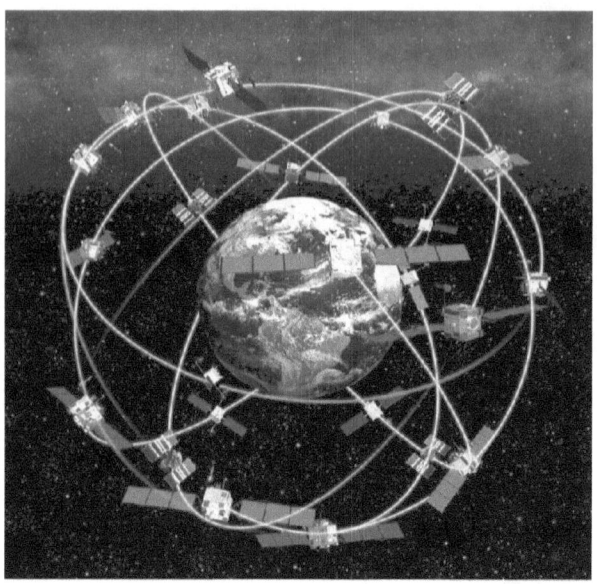

Instead of stars, we use satellites. Over 30 navigation satellites are zipping around high above Earth.

These satellites can tell us exactly where we are.

The Global Positioning System (GPS) is made up of satellites, ground stations, and receivers.

GPS is a system. It's made up of three parts: satellites, ground stations, and receivers.

Satellites act like the stars in constellations—we know where they are supposed to be at any given time.

The ground stations use radar to make sure they are actually where we think they are.

How high do the GPS satellites orbit?

There are currently 32 GPS satellites Orbiting earth at 12,427 miles above sea level

There are also 24 GLONASS satellites (The Russian version of a GPS satellite) Which many

High end GPS receivers will use in conjunction with the standard GPS satellites.

Chapter 73.1 Time Dilation Due to Speed

The orbital velocity from Newtonian Physics is 12,584 ft/sec.

Time duration passes faster at distances away from the surface of the earth in accordance

with special relativity. We use the Lorentz Transformation to calculate the duration of time difference between a clock on the surface of the earth and a satellite in orbit.

$$\text{Time (orbit)} = T \text{(earth)} / (1 - V^2/C^2)^{1/2}$$

C = velocity of light = 186,000 miles /second

V = orbital velocity = 2.3833 miles /second

T (orbit)/T (earth) = 1.000000001 seconds or 0.999999999 cps

The resonance frequency of a cesium atomic clock is 9,192,631,770 cycles/second on the earth. In orbit the clock may resonate at 9,192,631,779/ 0.999999999 cycles/second.

9,192.631,779 cps. The difference is 9 cps from 9,192,631,770 and the % error is 9.79 × 10⁻⁸ %.

The 9 cps is small but has been measured and confirmed for special relativity.

Chapter 73.2 Time Dilation due to Gravity

General Relativity states that gravity slows down the duration of time. Clocks in orbit keep faster time than clocks on earth. Time stands still in a black hole that has massive gravity.

The time dilation is as follows:

$$\text{Time (orbit)} = \text{Time (earth)} / (1 - 2 GM/RC^2)^{1/2}$$

M = is the mass of the earth

G = is the Universal Gravitational Constant

C = the velocity of light

R = is the distance from the center of mass (in this case the earth) = 12427 + 3900 miles

R = 16,327 miles

g = 32.2 ft/sec² (acceleration of gravity on surface of earth)

From Newton's Law of Gravity, $GM = (R + h)2\, g = 2.392 \times 10^{17}$ ft³/sec²

$RC2 = 8.314 \times 10^{25}$ ft³/sec²

$2\,GM/RC^2 = 2.877 \times 10^{-9}$

Time (orbit) = Time earth/0.999999985

or Orbital time = 1.000000015 × Time on earth

The resonance frequency of a cesium atomic clock is 9,192,631,770 cycles/second on the earth. In orbit the clock may resonate at 9,192,631,779/0.999999985 cycles/second or 9,192.631,917 cps.

The difference is 138 cps from 9,192,631,770 and the % error is 1.50×10^{-6} %.

The 138 cps is small but has been measured and confirmed for General Relativity

Chapter 73.3 Effect of Speed and Gravity on GPS Accuracy

The following is a summary from Wikipedia "Error Analysis for the GPS for Special and General Relativity"

The satellite clocks loses 7,214 × 10-9 (nanoseconds) /day due to Special Relativity

The satellite clocks gain 38,640 × 10-9 (nanoseconds)/ day due to General relativity

The clocks on the satellites have to be slowed down as follows:

From 10.23000000000 × 10^6 to 10.22999999543 × 10^6 (megahertz) to negate the effects of General Relativity.

There are other sources of errors such as cloud cover, snow, reflections of the GPS signals.

The accuracy of horizontal locations is about 50 ft +/- 16 feet.

The GPS system was created during the 1970s by the Dept of Defense for military purposes.

The technology was later commercialized and became available for automobiles.

Chapter 74.0 Newton's Laws

Isaac Newton formulated three two major branches of physics and one of mathematics, all of which transformed the worlds of science and math. He and Leibnitz conceived of the calculus. This book for high school students avoids using the calculus.

High School students should become familiar with Newton's Laws.

Three Laws of motion

First Law

A body in motion continues in motion unless acted upon by an external force. The external force may act as a resistance to slow the or change the direction of the body in motion or a force to speed up the body in motion. Forces to slow down a body in motion could be friction, air resistance, collisions, etc. Forces to increase motion could be pushing or pulling by humans or by a motor or engine of some kind.

Second Law

A force that acts on a body and causes it to move equals the product of the mass of the Body times the acceleration. However forces may be dynamic (causing motion) and some forces may be static (causing no motion). A force is defined as a push or a pull.

Examples of static forces are: Standing on a floor. The object pushed down on the floor with its weight and the floor pushes back with a reactive force equal to the weight.

Examples of dynamic forces Are forces that move buses, trains, and airplanes. Some forces are needed to overcome resistance to motion and may result in the object moving at a constant velocity. Automobiles drive at a constant velocity because the engine is sufficient to overcome friction between the ground and the tires and to overcome air resistance.

Acceleration is a measured quantity and is usually expressed as a change in velocity during a period of time. Velocity is usually expressed as feet/second or meters/second.

Acceleration or change in velocity is usually expressed as feet/second^2 or meters/second^2.

Force is a measured quantity and may be expressed as pounds or newtons.

One newton × 0.44 = 1 pound.

Mass in Newton's Equation was an invented concept by Newton and cannot be measured, only inferred from the Equation F=m×a, which is Newton's second Law. There is no such thing as a mass meter. Therefore mass can be expressed as F/a or lbs/ft/sec^2 or newtons/meter/sec^2. Use of these terms provide for dimensional consistency.

Engineers sometimes call 1.0 lbs/ft/sec^2 = one slug. Scientists sometimes call newton/meter/sec^2 = one kilogram. Sometimes kilograms are used for weight or for mass and this can be confusing. A mass (or material body) on th Earth has weight that can be measured. The weigh to mass on the Earth equals the downward force exerted by gravity on the mass. Falling bodies that were examined by Galileo is and measured to be 32.2 ft/sec^2 at sea level. Since F = m × a, we have F = m × 32.2 ft/sec^2 at sea level. A mass near the surface of the earth subject to gravity is called gravitational mass. The same mass elsewhere or pushed horizontally by a force is called inertial mass. Albert Einstein was puzzled by the fact that the inertial mass and gravitational mass are equal except when the mass is travelling at velocities close to the speed of light.

The inertial mass increases in accordance with the Lorenz Transformation.

Third Law

For every force acting on a mass, there is an equal and opposite force acting on the mass.

This means that forces come in pairs. The weight or force of an individual standing on a floor is opposed by an equal force of the floor pushing upwards. In the absence of an upward or counter force, then the mass accelerates in accordance with F = m ×a.

Chapter 74.1 Newton's Law of Gravity

Newton surmised that the invisible force that kept the planets in orbit around the sun was the same force that directed all untethered masses to fall to the earth. He must have used the data from Tycho Brahe and the calculations by Johaness Kepler to formulate that the force of attraction between the sun and planets is inversely proportional to the square of the distance between the bodies not inversely proportional to the distance between the bodies. Newton formulated the Law of Gravitational Attraction as a force:

$$F = \{M_e \times M_o \times G\}/d^2$$

F is force measured in ~ lbs. or newtons.

Me is the mass of the earth ~ kilograms or slugs

Mo is the mass of the object ~ in kilograms or slugs

G is the universal Gravitational Constant ~ 6.674×10^{-11} N·kg^{-2}·m^2.

d is the distance between the centers of both objects. ~ in feet or kilometers

The force or weight of an object at sea level + $m_o g$ where g = local acceleration of gravity

Therefore $F = \{M_e \times M_o \times G\}/ D^2 = M_o g$

or at sea level $Me \times G = g \times D^2$

At sea level d = earth radius.

UNTAUGHT MATH

Chapter 74.2 Centrifugal and centripetal forces.

Newton also showed that an object in orbit has two forces acting on it.

Centrifugal force tends to pulls the object out of the orbit.

Centripetal force pulls the object toward the larger mass. This is gravity.

Gravitational force = centrifugal force. For an object in orbit.

Newton showed that the centrifugal force = mass × velocity² / distance. Or

Centrifugal force = M V² /Distance

M is the mass of the object in orbit

V is the velocity of the object in circular motion

$M V^2 /R = M g$ $V^2 = R g$

R is the radius of the earth

These are some of the fundamental relationships we learned from Isaac Newton.

Centrifugal Force

While the earth is spinning, why are we not thrown off by the centrifugal force?

Centrifugal force is balanced by the force of gravity.

The strongest centrifugal force is at the equator of the earth at a radius of 3900 miles.

Let us examine the ratio of the force of gravity divided by the centrifugal force.

Centrifugal Force = Mass × Velocity^2/ Distance from center of the earth.

Gravitational force at sea level = Mass × Acceleration of gravity at sea level.

Gravitational force/ Centrifugal Force = M g/ M V^2 / R = g × R/V^2

M = mass of the object

g = 32.2 ft/sec^2

R = 3900 miles = 2.059 × 10^7 feet

V = tangential velocity at the equator = Circumference of the earth/ time per revolution

= 2 × 3.14 × 3900 × 5280 = 1.292 × 10^8 feet = circumference of the earth

Time per rotation of the earth = 24 hours × 3600 seconds/hour = 8.64 × 10^4 seconds

Tangential Velocity at the equator = 1.495 × 10^3 feet/second

V^2 = 2.236 × 10^6 ft^2/sec^2

Ratio = 32.2 × 2.059 × 10^7/2.236 × 10^6 = 296

The maximum centrifugal force is at the equator = about 0.034% of the gravitational force.

That is why we are not thrown off the earth by centrifugal force that is a result of the Spinning of the earth. The earth would have to spin 17 times faster for the gravitational force to equal the centrifugal force. One rotation would last 1.4 hours instead of 24 hours.

Chapter 75.0 Growth Models

Time Minutes	Linear Growth $G = KT$	Accelerated Growth $G = G_0 T^2$	Exponential Growth $G = G_0 e^{KT}$
0	1.0	1.0	1.0
1	2.0	4.0	2.72
2	3.0	9.0	7.39
3	4.0	16.0	20.08
4	5.0	25.0	54.60
5	6.0	36.0	148.41

$G_0 = 1.0$ $K = 1.0$ Decay patterns are obtained when G_0 and/or K are negative.

Growth patterns are described as linear, accelerated or exponential.

Chapter 76.0 Probability

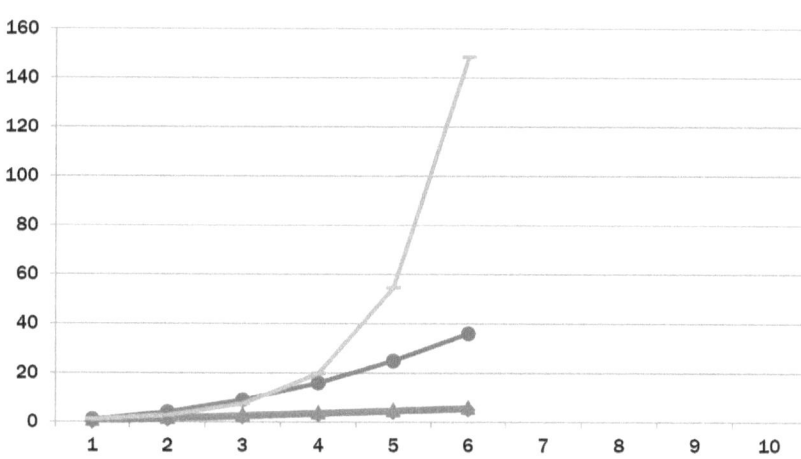

Chapter 76.1 Birthday Problem (From Wikipedia)

What is the probability that two people have identical birthdays within a group of N people?

Solution is: $P(A) = \dfrac{365}{365} \times \dfrac{364}{365} \times \dfrac{363}{365} \times \dfrac{362}{365} \times \ldots\ldots\ldots \dfrac{343}{365}$

365 - 343 = 22 people

The solution is: Evaluating equation gives $P(A) \approx 0.492703 = 50\%$

Therefore, $P(A) \approx 1 - 0.492703 = 0.507297$ (50.7297%). = 50%

In other words, in a group of 22 people the probability that two people have the same birthday is 50%.

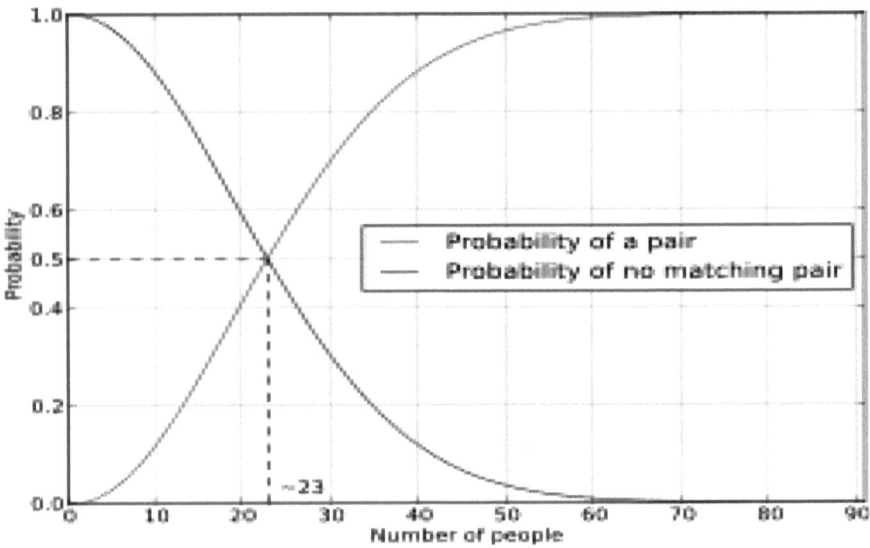

Chapter 76.2 Probability of Achieving Various Hands at Poker

Probability is defined as the chance of success divided by the total opportunities. The calculations below assume that you have the first opportunity to select cards from a deck of 52 cards otherwise there are fewer cards in the deck.

Royal Flush: 10, Jack, Queen, King, Ace—all of the same color

$4/52 \times 1/51 \times 1/50 \times 1/49 \times 1/48 \times 100 = 2.16 \times 10^{-8}\%$

Straight Flush: 10, Jack, Queen, King, Ace—all of different color

$2.16 \times 10^{-8}\% \times 4^5 = 2.22 \times 10^{-5}\%$

Three of a kind—color of cards is irrelevant

$52/52 \times 3/51 \times 2/50 \times 100\% = 0.235\%$

Full House: Example: QQQ33 or 66699

$52/52 \times 3/51 \times 2/50 \times 49/49 \times 3/48 \times 100\% = 0.0147\%$

Two of a kind

$52/52 \times 3/51 \times 100\% = 5.88\%$

Chapter 76.3 Statistics

A basic understanding of statistics is essential for a liberal education. Statistics are used in many disciplines in colleges: psychology, sociology, business planning, etc. In this case we shall examine how to describe the data about height for males and females using the acceptable mathematical parameters of mean, mode, median, standard deviation, range, population and normal curve.

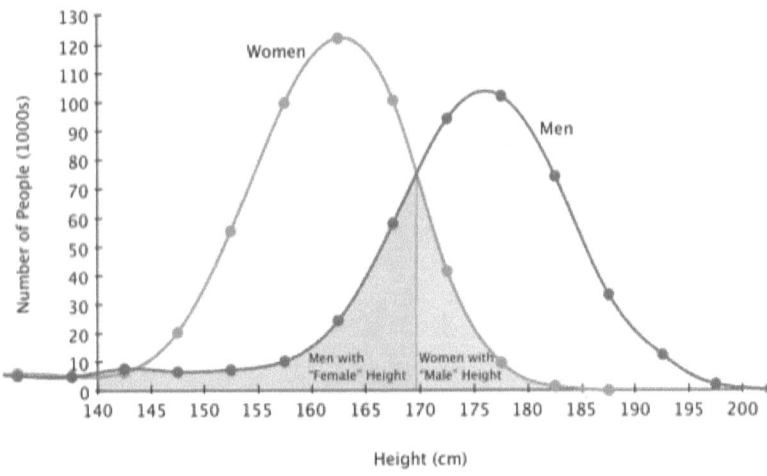

Height (cm)

Normal curve: A normal curve or normal distribution looks like a "Bell Shaped Curve." This is sometimes called a Gaussian Distribution named after Carl Friedrich Gauss. The curve has a maximum value usually in the middle and asymptotically approaches zero on the ends range: The total horizontal or X axis is called the range, and for women varies from less than 140 centimeters to 187 centimeters (<4.6 feet to 6.1 feet) For men the range varies from less than 140 centimeters to 205 centimeters. (< 4.6 feet to 6.7 feet).

Population: The population or sample size for women appears to be 125,000 and 100,000 for men. The population is plotted on the vertical or Y axis.

Mean: The mean or average height for women appears in the middle of the symmetric Bell Curve and is 162.5 centimeters (5.33 feet). The mean or average height for men appears in the middle of the symmetric Bell Curve and is 176,5 centimeters (5.79 feet).

Median: The median is defined as the height at the 50% point on the X axis for men and for women calculated as follows: Median = Range/2 + minimum value = $\frac{(205 - 140)}{2} + 140 = 172.5$ centimeters. For women it is $\frac{(187 - 140)}{2} + 140 = 163.5$ centimeters.

For a perfect Bell Shaped curve the medians are equal to the means.

Mode: A mode is a value (in a set of values) which is more common than other values in that set. Conventional statistics commonly assumes data will have a single mode, close to its midrange which is what we have in the Bell Shaped Curve. Assume our data is presented as follows: 140, 150, 150, 150, 160, 160, 170, 180. Then the mode is the most frequently occurring number and is 150. Mode calculations are applicable if the data is presented in packages.

Standard Deviation: The standard deviation measures the spread of the data on the Bell Shaped Curve. Data very close to the mean has a small standard deviation.

The Greek letter sigma σ is used to denote the standard deviation for large populations:

$$\text{Sigma} = \sqrt{\frac{(\text{Mean} - \text{Height})^2}{N}} \quad \text{for each height, H}$$

$(162.5 - 140)^2 = 506.25$

$(162.5 - 150)^2 = 156.25$

$(162.5 - 155)^2 = 56.25$

$(162.5 - 160)^2 = 6.25$

$(162.5 - 165)^2 = 6.25$

$(162.5 - 170)^2 = 56.25$

$(162.5 - 175)^2 = 156.25$

$(162.5 - 180)^2 = 306.25$

$(162.5 - 185)^2 = 506.25$

$(162.5 - 187)^2 = 600.25$

Sum $= 2356.5 = \sigma = \sqrt{\dfrac{2356.5}{10}} = 15.35 =$ One standard deviation.

The difference between three and six sigma is what percentage of the observations that make up the total sample or population will fall between the sigma level and the mean. The rule of "68-95-97" states that approximately 68 percent of the observations in a normally distributed sample fall within plus or minus one standard deviation from the mean, while 95 percent fall within two standard deviations, and just over 97 percent fall within three standard deviations. For the example above, 68% of the area under the Bell Curve includes the heights between the mean +/- 15.35 = 162.5 - 15.35 to 162.5 + 15.35. or 147.15 to 177.85

1. $\sigma = 68\%$ The height of 68% of females are between 147 to 177 centimeters
2. $\sigma = 95\%$ The height of 95% of females are between 132 to 193 centimeters

3. σ = 97% The height of 97% of females are between 117 to 208 centimeters.

A good example would be to look at the normal distribution (this is not the only possible distribution though).

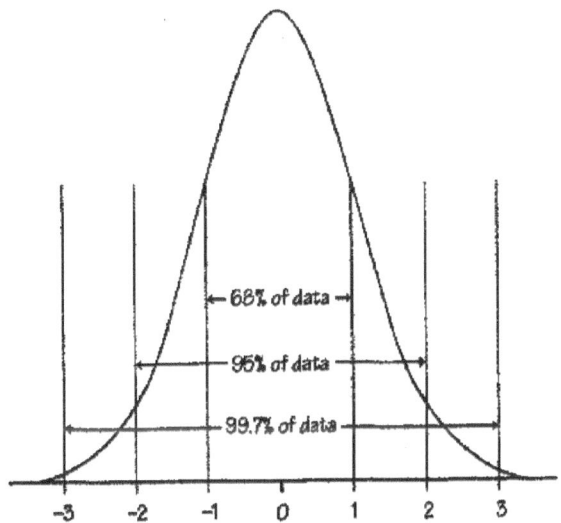

In this image, 0 would be the mean, or 0 standard deviations from the mean. 1 would be 1 standard deviation from the mean.

Chapter 77.0 Set Theory

George Cantor founded Set Theory and published his findings in 1874. A set consists of a group of elements that may or may not have anything in common except that the elements belong to the same set.

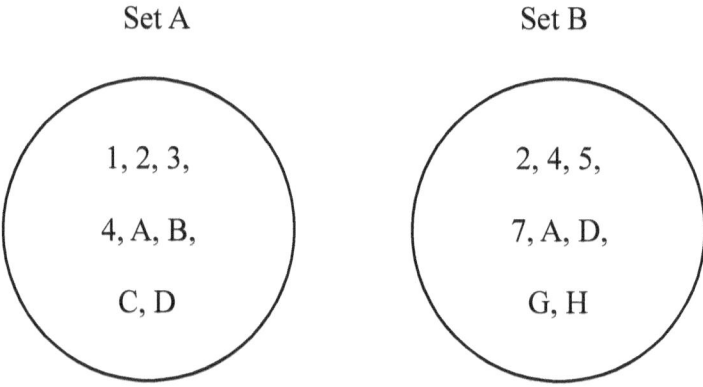

The Union of Sets A & B consists of all members of A and B or both A U B = {1, 2, 3, 4, 5, 7, A, B, C, D, G, H}

The Intersection of A and B consists of only the elements common to both sets.

A ∩ B = {2, 4, A, D}

The set difference A \ B consists of elements of A not in B A \ B = {1, 3, C}

The set difference B \ A consists of elements of B not in A. B \ A = {5, 7, G, H}

Note that: A \ B does not equal (≠) B \ A

The null set has no elements and is denoted by: φ

A ∩ φ = 0

A U φ = A

Chapter 78.0 Equal Product and Sum of Two Integers

We are asked to find two integers whose sum and product are equal.

Let X = Integer 1   Let Y = Integer 2   Let N = Sum and Product

X Y = N and X + Y = N

Substitution: X (N - X) = N or $X^2 - NX - N = 0$

N >= 4 can be any integer, Let N = 10 for this example

$X = \frac{\pm N \pm \sqrt{N^2 - 4N}}{2}$ If N = 10;   X = ½(10 + 7.745) = 8.872

Y = N - X = 10.0 - 8.872 = 1.127   XY = 9.9989   X + Y = 9.999

If N = 10 X also = ½ (10.0 - 7.745) = 1.1275 and Y = 8.8725

XY = 10.0037, X + Y = 10.000

If N = 4; then Y = X = 2

Chapter 78.1 "Eight This Cool" (Jewish Home Magazine) Page 82, 6 Dec 2018

$9 \times 9 + 7 = 88$

$98 \times 9 + 6 = 888$

$987 \times 9 + 5 = 8888$

$9876 \times 9 + 4 = 88888$

$98765 \times 9 + 3 = 888888$

$987654 \times 9 + 2 = 8888888$

$9876543 \times 9 + 1 = 88888888$

$98765432 \times 9 + 0 = 888888888$

Chapter 78.2 Cos (Cos (Cos (Cos Θ))) etc. = 0.7389

Limit as i ⟶ 00, Cos (Cos (Cos (Cos Θ))) etc.,. ⟶ 0.7389, Θ is in radians

Limit as i ⟶ 00, Sin (Sin (Sin (Sin Θ))) etc.,. ⟶ 0.0, Θ is in radians

	cos 1.0	cos 0.5	Cos 0.01	Sin 1.0
1	0.540302	0.87758	0.99995	0.841471
2	0.857553	0.639014	0.540344	0.745624
3	0.65429	0.802684	0.857532	0.67843

4	0.79348	0.694779	0.654306	0.627572
5	0.701369	0.768195	0.79347	0.587181
6	0.76396	0.719166	0.701376	0.554016
7	0.722102	0.752356	0.763955	0.526107
8	0.750418	0.730081	0.722106	0.502171
9	0.731404	0.74512	0.750416	0.481329
10	0.744237	0.735006	0.731405	0.462958
11	0.735605	0.741826	0.744236	0.446597
12	0.741425	0.737236	0.735605	0.431898
13	0.737507	0.74033	0.741425	0.418596
14	0.740147	0.738246	0.737507	0.406478
15	0.738369	0.73965	0.740147	0.395377
16	0.739567	0.738705	0.738369	0.385156
17	0.73876	0.739341	0.739567	0.375703
18	0.739304	0.738912	0.73876	0.366927
19	0.738938	0.739201	0.739304	0.358749
20	0.739184	0.739007	0.738938	0.351103
21	0.739018	0.739138	0.739184	0.343934
22	0.73913	0.73905	0.739018	0.337193
23	0.739055	0.739109	0.73913	0.330839
24	0.739106	0.739069	0.739055	0.324837
25	0.739071	0.739096	0.739106	0.319154
26	0.739094	0.739078	0.739071	0.313764
27	0.739079	0.73909	0.739094	0.308641
28	0.739089	0.739082	0.739079	0.303764
29	0.739082	0.739087	0.739089	0.299114
30	0.739087	0.739084	0.739082	0.294674
1000				0.054839
1200				0.054839
2000				0.038749
7000				0.020704
10000				0.017321

Chapter 78.3 Asymptotes

An asymptote, for an open-ended polynomial on a graph, shows the edges of the function displayed on a graph

Example 1. Consider the open ended hyperbolic function $Y = 1/X$

A table for the function is a s follows:

X	Y
1.0	1.0
-5	- 0.20
+5	+ 0.20
-4	- 0.25
4	+ 0.25
-3	- 0.33
+3	- 0.33
-0.5	- 2.00
0.5	+ 2.00
-0.1	- 10.0
+ 0.1	+ 10.0
+10.0	+ 0.10
-10.00	- 0.10

We can see from the graph as $X \longrightarrow +/- \infty$ (infinity); $Y \longrightarrow 0$

Similarly, as $Y \longrightarrow +/- \infty$, $X \longrightarrow 0$. The asymptotes are $X = 0$ and $Y = 0$.

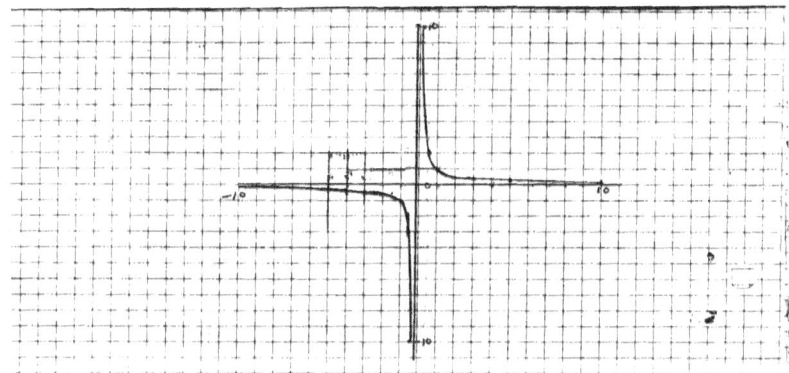

Example 2 Consider the open ended hyperbolic function $Y = 1/(X-1)$

A table for the function is a s follows;

X	Y
10.0	0.11
-10.0	-0.909
8.0	0.142
-8.0	-0.111
6.0	0.200
-6.0	-0.142
4.0	0.333
-4.0	-0.200
2.0	1.00
-2.0	-0.333
1.00	∞
-1.00	-0.500
0.50	-2.00
-0.50	-0.666
0.20	-1.25

-0.20	-0.833
11.0	0.100
-11.0	-0.083
0.10	-1.10
-0.10	-0.909
0.10	- 1.11
3.0	0.50

The asymptote for X may be obtained by setting the denominator = 0.

As X ⟶ 1.0   Y ⟶ ∞

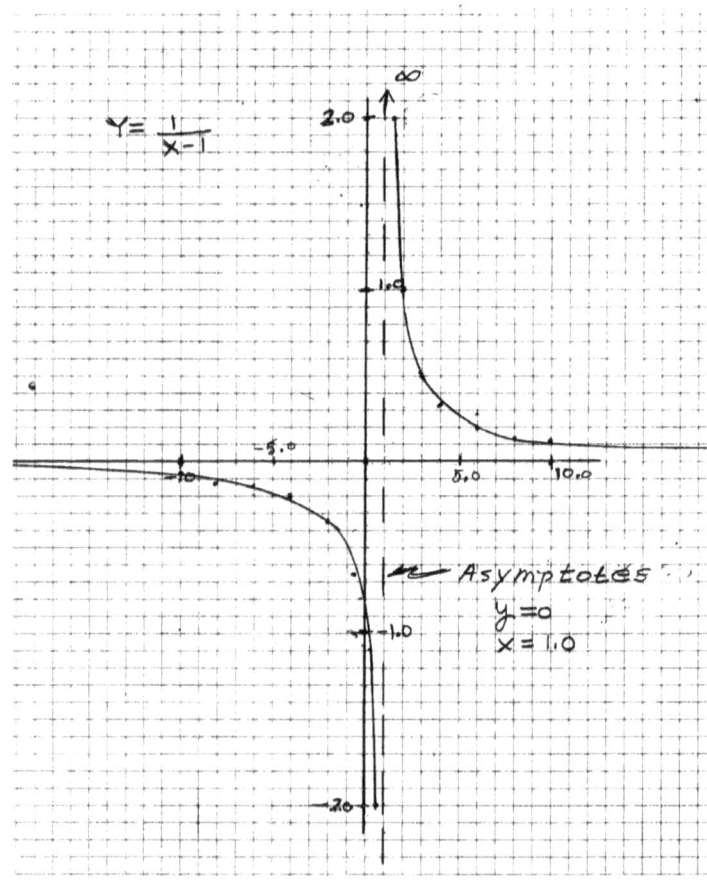

Example 3 $Y = -2X^3 / (x + 1)$

The asymptote may be found by setting the denominator = 0.
Asymptote, $X = -1$. As $X \longrightarrow -1.0$, $Y \longrightarrow \infty$

$$-2\,X\,X\,X / (X+1)$$

X	Y
0	0
0.1	-0.00182
0.2	-0.01333
0.3	-0.04154
0.4	-0.09143
0.5	-0.16667
0.6	-0.27
0.7	-0.40353
0.8	-0.56889
0.9	-0.76737
1	-1
0.1	0.002222
0.2	0.02
0.3	0.077143
0.4	0.213333
0.5	0.5
0.6	1.08
0.7	2.286667
0.8	5.12
0.9	14.58
-1	#DIV/0!
1.1	-1.26762
1.2	-1.57091

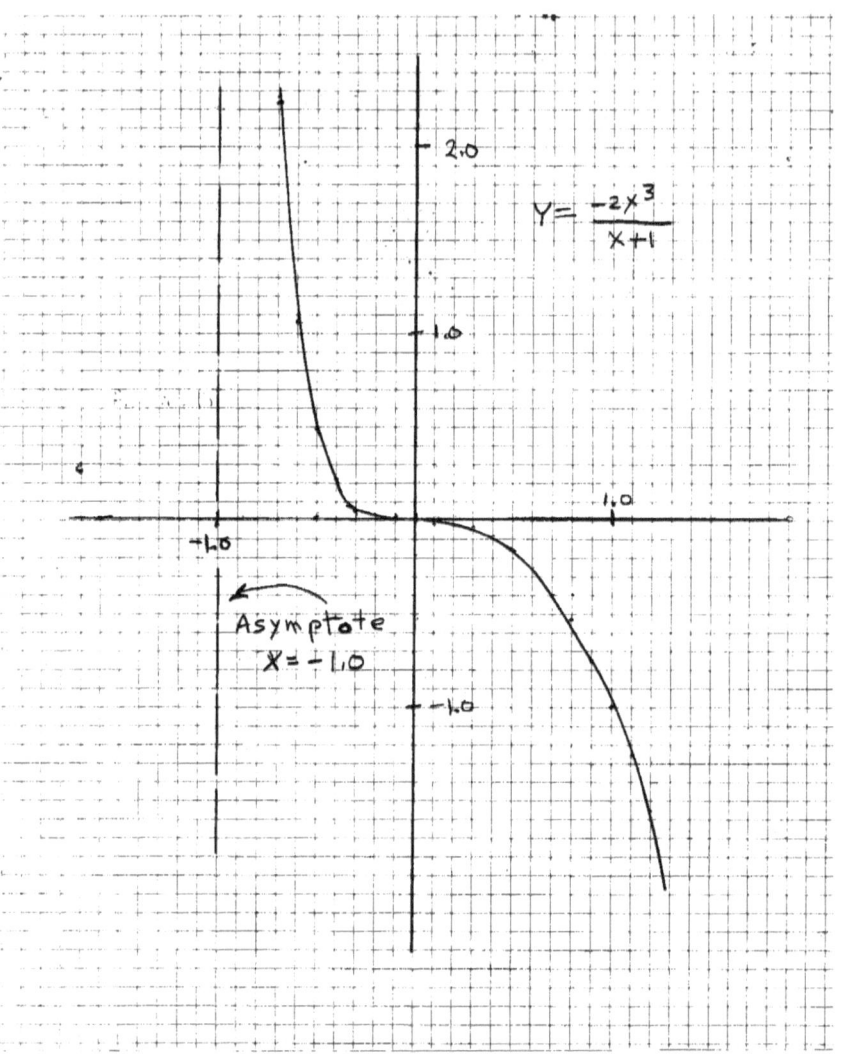

Example 4   Y = (X. X -3 X - 2)/(X - 2)

This is an example of a side asymptote.

Y = (X. X -3 X - 2)/(X - 2)

X	Y
0	1.00
0.2	1.42
0.4	1.90
0.6	2.46
0.8	3.13
1	4.00
-0.2	0.62
-0.4	0.27
-0.6	-0.06
-0.8	-0.37
-1	-0.67
1.2	5.20
1.4	7.07
1.6	10.60
1.8	20.80
2	#DIV/0!
-1.2	-0.95
-1.4	-1.22
-1.6	-1.49
-1.8	-1.75
-2	-2.00
-2.2	-2.25
-2.4	-2.49
-2.6	-2.73
-2.8	-2.97

-3	-3.20		
2.2	-18.80		
2.4	-8.60		
2.6	-5.07		
2.8	-3.20		
3	-2.00		
3.2	-1.13		
3.4	-0.46	-3.8	-4.11
3.6	0.10	-4	-4.33
3.8	0.58	-4.2	-4.55
4	1.00	-4.4	-4.78
4.2	1.38	-4.6	-4.99
4.4	1.73	-4.8	-5.21
4.6	2.06	-5	-5.43
4.8	2.37	5.2	2.95
5	2.67	5.4	3.22
-3.2	-3.43	5.6	3.49
-3.4	-3.66	5.8	3.75
-3.6	-3.89	6	4.00

How to compute slant asymptotes

$Y = (X^2 - 3X - 2) / (X - 2)$

As $X \longrightarrow 2$; $Y \longrightarrow \infty$ (infinity) therefore $X = 2$ is an asymptote.

The slant asymptote equation is obtained by dividing $(X^2 - 3X - 2)$ by $X - 2$ using synthetic division as follows:

```
           X - 1
       ┌─────────────
X - 2  │ X² - 3X - 2
         X² - 2X
         ─────────
              -X - 2
              -X + 2
              ──────
                  - 4
```

(X² - 3X - 2)/(X - 2) = X - 1 - 4 / (X – 2)

Y = X - 1 is the slant asymptote.

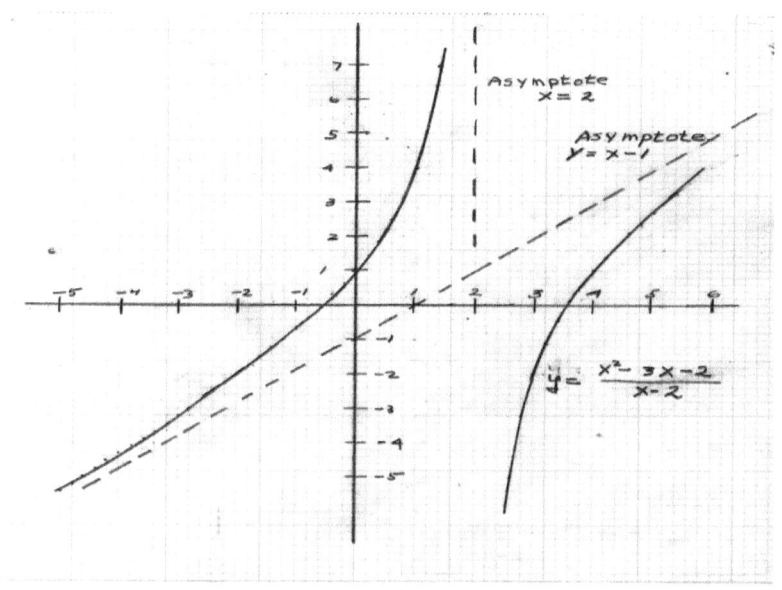

Chapter 78.4 Asymptotes for a Hyperbola

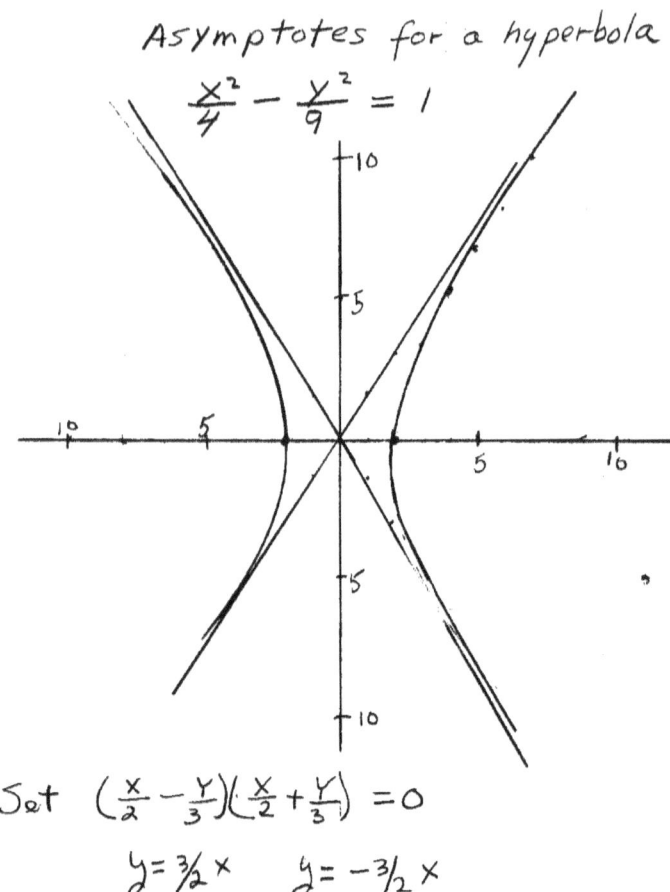

Asymptotes for a hyperbola

$$\frac{x^2}{4} - \frac{y^2}{9} = 1$$

Set $\left(\frac{x}{2} - \frac{y}{3}\right)\left(\frac{x}{2} + \frac{y}{3}\right) = 0$

$y = \frac{3}{2}x \qquad y = -\frac{3}{2}x$

Chapter 79.0 Finding the Areas of SAS, ASA, and SSS Triangles.

SAS Triangle means that the triangle is defined by two sides and the included angle.

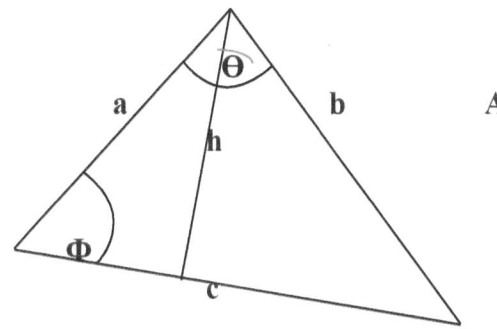

Area = ½ base X height

Angle Θ is between a and b and is known, a and b are known.

Drop a perpendicular line called "h" to be the altitude of the triangle.

Use the Law of Cosines to find side c as follows:

$c^2 = a^2 + b^2 - 2ab \cos \Theta$; solve for c; c is the base

Now use the Law of Sines to find angle Φ as follows:

$c / \sin \Theta = b / \sin \Phi$ Solve for angle Φ

Now solve for h $h/a = \sin \Phi$

We now know base = c and height = h

Area = ½ base × height

Now solve the area for angle-side-angle (ASA)

We know Θ and Φ and side a.

Find side c by using the Law of Sines

c / Sin Θ = a / Sin (180 - Θ - Φ)

Now solve for h h / a = Sin Φ; we now know base and height

Area = ½ base × height

Now solve the area if a, b, and c are known.

Use the Law of Cosines to find angle Θ

$c^2 = a^2 + b^2 - 2 a b \cos \Theta$; solve for Θ; c is the base

Use the Law of Sines to find Φ

C / Sin Θ = b / Sin Φ

Now solve for h h / a = Sin Φ; we now know base and height

Area = ½ base × height

Chapter 80.0 Completing the Square for Solving Quadric Equations

Completing the square for solving quadratic equations and deriving the standard solution

We have a quadratic equation: $ax^2 + bx + C = 0$

Divide terms by a. $x^2 + bx/a + c/a = 0$

Create two equal factors: $(x + b\,x/2\,a)^2 + c/a - (b/2\,a)^2 = 0$.

The expansion above reduces to the same quadratic equation.

Rewriting $(x + b/2\,a)^2 = (b/2\,a)^2 - c/a$

Take the square root of both sides: $(x + b/2\,a) = \sqrt{(b/2\,a)^2 - c/a}$

Solve for x: $x = -b/2\,a +/- (1/2)\sqrt{b^2 - 4ac}$

Finally we get: $x = \dfrac{-b \pm \sqrt{b^2 - 4ac}}{2\,a}$

Chapter 81.0 Length of a Chord

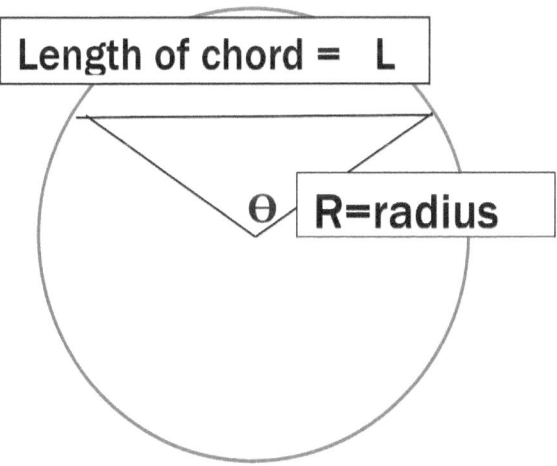

Use the Law of Cosines

$L^2 = R^2 + R^2 - 2R^2 \cos \Theta$

$L = R [2(1 - \cos \Theta)]^{1/2}$

If $\Theta = 0$, then $L = 0$

If $\Theta = 180$, then $L = 2R$.

Chapter 81.1 Area of a Segment of a Circle

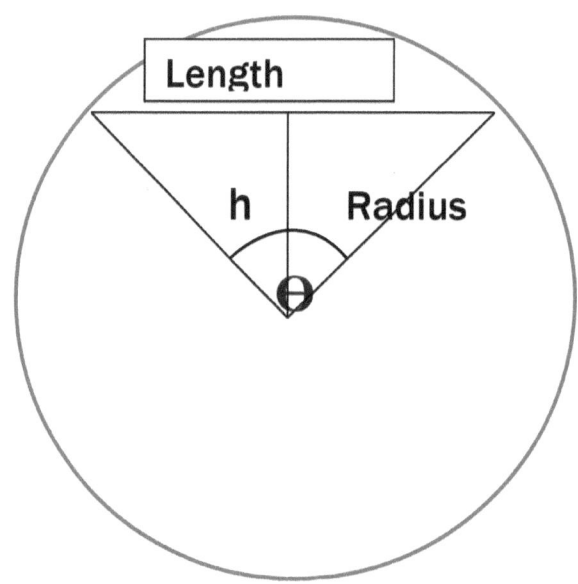

h = distance from center of circle to chord

Length = Length of Chord = L

Radius of circle = R

From Chapter 81.0 $L = R [2(1 - \cos \Theta)]^{1/2}$

h/R = Cos Θ/2

Area of the triangle = ½ L × h

$= R^2/2 \,[2\,(1 - \cos\Theta)]^{1/2} \cos\Theta/2$

Area of the segment = Area of the Sector - Area of the Triangle

$= R^2 \,\{\pi\Theta/360 - \frac{1}{2}\,[2\,(1 - \cos\Theta)]^{1/2} \cos\Theta/2\}$

Chapter 81.2 Centroid of a Triangle

There are many ways to prove that the centroid is 2/3 from the vertex of each median. The centroid is the center of gravity of the triangle.

In the figure below, angle φ is the top vertex of the triangle. M1 and M2 are medians. The medians bisect the opposite sides. We shall prove that the centroid is 2/3 from each vertex. We are using only 2 of the 3 centroids.

Statement

1. Triangles AEB and ADC are similar.
2. Therefore the ratio of DC/EB = 2:1
3. Angle a = Angle d
4. EB and Dc are parallel
5. Angle f = angle f

Proof

Angle φ is common to both triangles
The ratios of sides AD/AE and AC/AB = 2:1
DC and EB are legs of similar Triangles
AEB and ADC are similar
Alternate interior angles equal
Vertical angles are equal

6. EBF and DFC are   Sides DC and EB
   also similar   have ratios of 2:1

Sides DF/FB = 21   F is the intersection of
and CF/FE = 2:1   the Medians 2/3 from the
    respective verticies

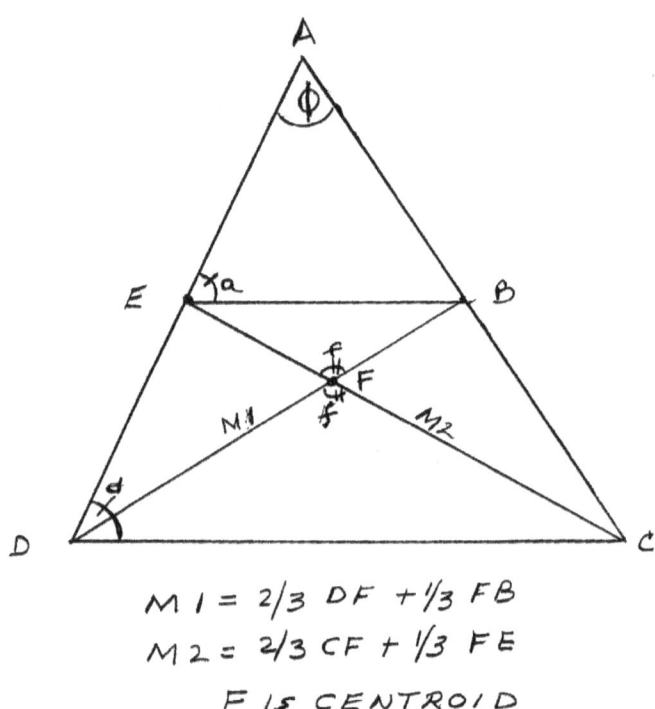

M1 = 2/3 DF + 1/3 FB
M2 = 2/3 CF + 1/3 FE
F IS CENTROID

Chapter 81.3 Length of Medians of a Triangle

Length of the Medians knowing the sides of the Triangle

Refer to Diagram in Chapter 81.2

Let AD = a

Let AC = b   AB = b/2

Let DC = c

Law of Cosines: $c^2 = a^2 + b^2 - 2ab \cos \phi$

$\cos \phi = (a^2 + b^2 - c^2)/2ab$

Use Law of Cosines again for M1 $M1^2 = a^2 + (b/2)^2 - 2a(b/2) \cos \phi$

Solving for M1 $M1^2 = \frac{1}{2}(a^2 - b^2/2 + c^2)$

Similarly solving for M2

$M2^2 = \frac{1}{2}(b^2 - c^2/2 + a^2)$

Solving for M3 (not shown) $M3^2 = \frac{1}{2}(c^2 - a^2/2 + b^2)$

Adding up all the Medians squared

$M1^2 + M2^2 + M3^2 = \frac{3}{4}(a^2 + b^2 + c^2)$

The sum of the squares of the medians = ¾ of the sum of the squares of the sides.

Chapter 81.4 Inequalities and Graphical Display

Inequalities Display

Expressions that display inequalities such as;

Y > 4 or X < -5 as well as inequalities for simultaneous equations may be expressed on a graph as shown below:

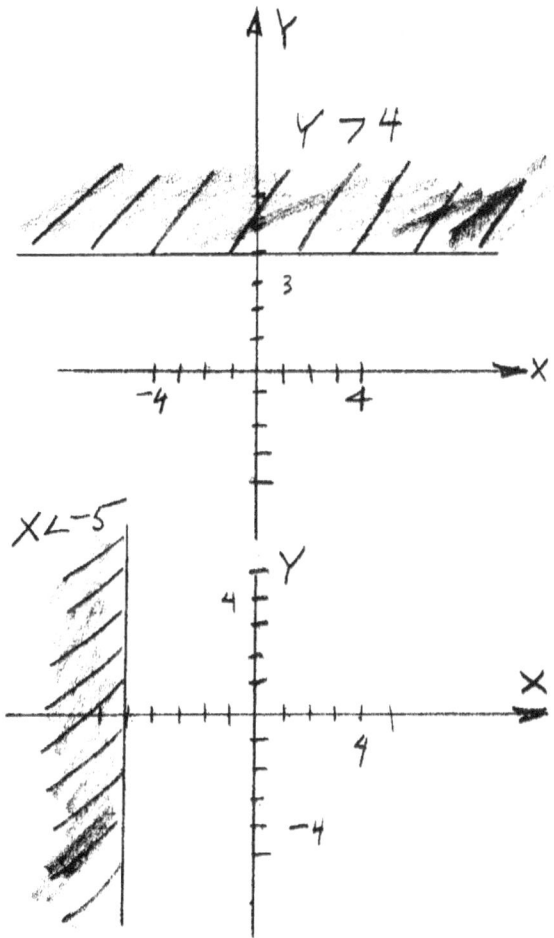

The two simultaneous inequalities below are t: Y < 2X + 4 & Y < -X/2 − 4, Y < 2 X + 4 & Y > -X/2 -4, Y > 2 X + 4 & Y < -X/2 -4 , Y > 2X +4 & Y > -X/2 - 4

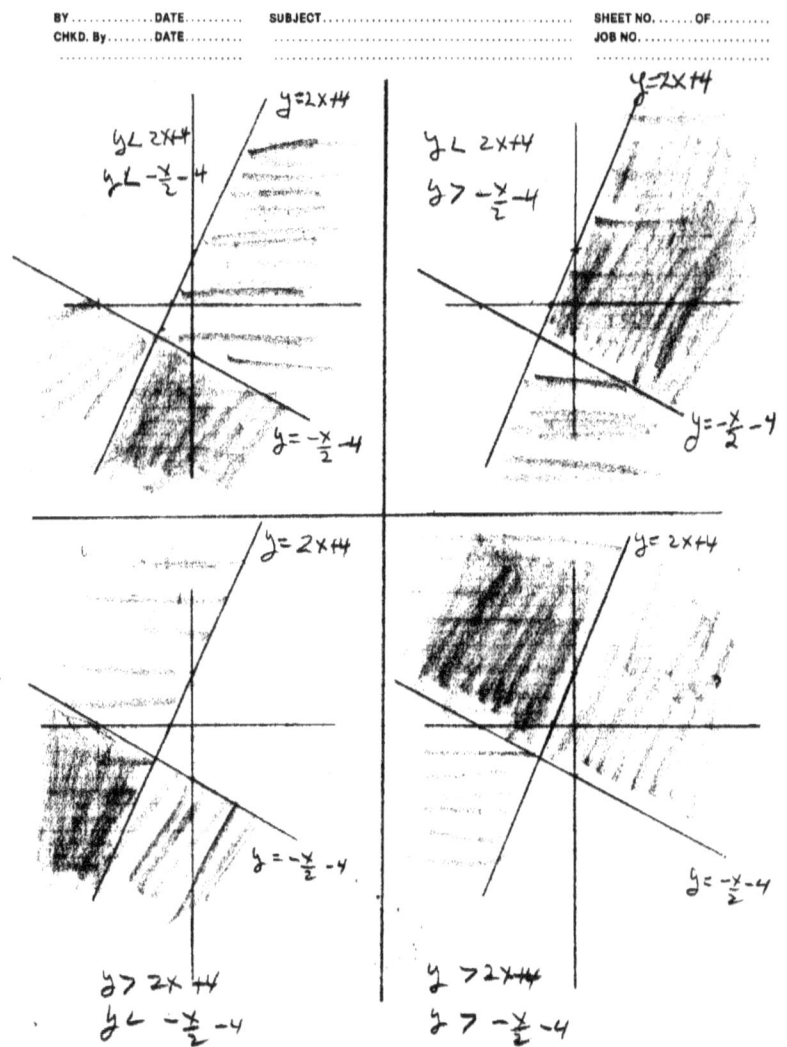

Chapter 81.5 Resultant of Adding Two Vectors Using Law of Cosines

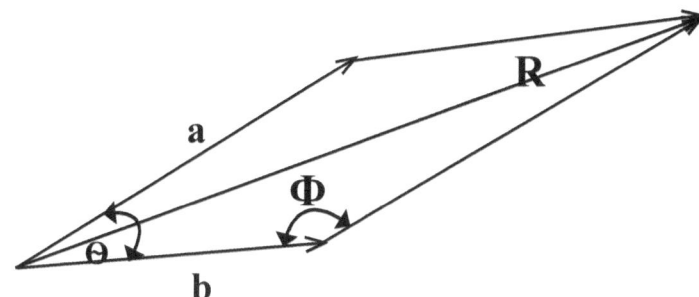

Θ is the angle between sides vectors a and b. Resultant vector is R. We want to find the magnitude of R.

We start with the parallelogram: Angle Θ + Angle Θ + Angle Φ + Angle φ = 360

Φ = 180 - Θ

R2 = a2 + b2 - 2 a b Cos φ

Cos Φ = Cos (180 - Θ) = Cos 180 Cos Θ + Sin 180 Sin Θ = - Cos Θ

Trigonometric Identities

Cos (A + B) = Cos A Cos B - Sin A Sin B

Sin (A + B) = Sin A Cos B + Cos A Sin B

Cos (A − B) = Cos A Cos B + Sin A Sin B

Sin (A − B) = Sin A Cos B - Cos A Sin B

Therefore we get: $R^2 = a^2 + b^2 + 2 a b \cos \Theta$

Chapter 81.6 Resultant of Subtracting Two Vectors Using Law of Cosines

$$R^2 = a^2 + b^2 - 2ab\cos\Theta$$

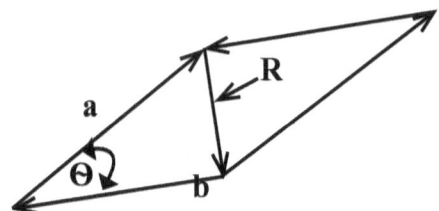

Chapter 81.7 Area and Perimeter of an Ellipse.

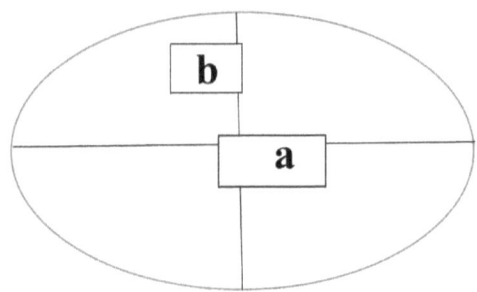

a is the major axis & b is the minor axis.

Equation of Ellipse $X^2/a^2 + Y^2/b^2 = 1$

Area = $\pi a b$ Circumference = $2\pi \sqrt{(a^2 + b^2)/2}$

Chapter 82.0 Collatz Problem

Consider the following operation on an arbitrary positive integer:

If the number is even, divide it by two. If the number is odd, triple it and add one.

Now repeat the operation.

The Collatz conjecture is: This process will eventually reach the number 1, regardless of which positive integer is chosen initially.

As an example chose 21, 21×3+1=64;64/2=32;32/2=16; 16/2 = 8 8/2 = 4; 4/2 = 2; 2/2 = 1

Chose 17: 17 × 3 + 1 = 52; 52/2 = 26; 26/2 = 13; 13 × 3 + 1 = 40; 40/2 = 20; 20/2 =10; 10/2 = 5; 5 × 3 + 1 = 16; 16/2 8; 8/2 = 4;4/2 = 2; 2/2 = 1

Chapter 83.0 Quadratic Equations

Quadratic Equation has the form: $f(X) = 0 = aX^2 + bX + C$

Divide all terms by a.

Then we get $X^2 + b/a\ X + C/a = 0$

Factor the equation as follows: $(X + b/2a)^2 + (b/2a)^2 + C/a = 0$

Since $(X + b/2a)^2 = X^2 + b/a\ X + (b/2a)^2 - (b/2a)^2 + c/a = 0$

$\sqrt{(X + b/2a)2} = \sqrt{(b/2a)^2 - 4a\ c/4aa}$

$X + b/2a = 1/2a \{ \sqrt{(b)2} - 4ac \}$

Let X1 and X 2 be the two roots

$$x = \frac{-b \pm \sqrt{(b^2-4ac)}}{2a}$$

$$X1 = \frac{-b+\sqrt{(b^2-4ac)}}{2a} \qquad X2 = \frac{-b-\sqrt{(b^2-4ac)}}{2a}$$

X 1 + X 2 = -b/a

X1 times X 2 = c/a

Therefore;

The sum of the roots = - b/a for any quadratic equation

The product of the roots = c/a for any quadratic equation

Chapter 83.1 Parabola

The form for the standard equation for a parabola is:

$(Y - H)^2 = 4P (Y - K)$

Definitions:

Vertex of the parabola is H, K on a grid

The latus rectum (straight side in Latin) = 2 P

The directrix = K - P

X Intercept is value of X when Y = 0

Y intercept is the value of Y when X = 0

If P > 0 then parabola opens upwards U

If P < 0 then [parabola opens downwards

Start with the traditional form of a quadratic = Y as follows:

$Y = aX^2 + bX + C$

Divide by a

$Y/a = (X + b/2a)^2 - (b/2a)^2 + c/a$

$(X + b/2a)^2 = Y/a + (b/2a)^2 - c/a$

$(X + b/2a)^2 = Y/a - \{c/a - (b/2a)^2\}$

4P = 1.0 P = 1/4

Length of talus rectum (means straight side) = 2P

Vertex (H, K) on a graph which equals (b/2a, c/a - (b/2a)^2

Directrix on the graph = K - P = c/a - (b/2a)^2

Focus: (H , P + K) = (H/2a, ¼ + c/a - (b/2a)2)

Y intercept (when X = 0) = c

X intercept (when Y = 0)

$$x = \frac{-b \pm \sqrt{(b^2-4ac)}}{2a}$$

Now we shall use an example of a parabola

$X^2 - 10X - 23 = 12Y$

$(X - 5)^2 - 25 - 23 = 12Y$

$(X - 5)^2 = 12Y + 48$

Standard Form of a parabola $(X - H)^2 = 4P(Y - K)$

$(X - H)^2 = 4PY - 4PK$; $4PY = 12Y$ & $-4PK = 48$

Therefore $H = 5$ and $P = 3$ and $K = -4$

Vertex: $(5, -4)$ Directrix: $= K - P = -7$

Focus $= \{H, P + K\} = \{5, -1\}$

Y intercept $(X = 0)$ $Y = -23/12 = -1.916$

X intercept $(Y = 0)$ $X =$

$$x = \frac{-b \pm \sqrt{(b^2 - 4ac)}}{2a}$$

$a = 1.0$, $b = -10$, $c = -23$

$X = +11.9$ and $X = -1.90$

Minimum value for parabola $X = 5$ and $Y = -4.0$ (using calculus by taking derivative $dy/dx = 0$)

Latus rectum $= 2P = 6.0$

Now we show the parabola on a graph

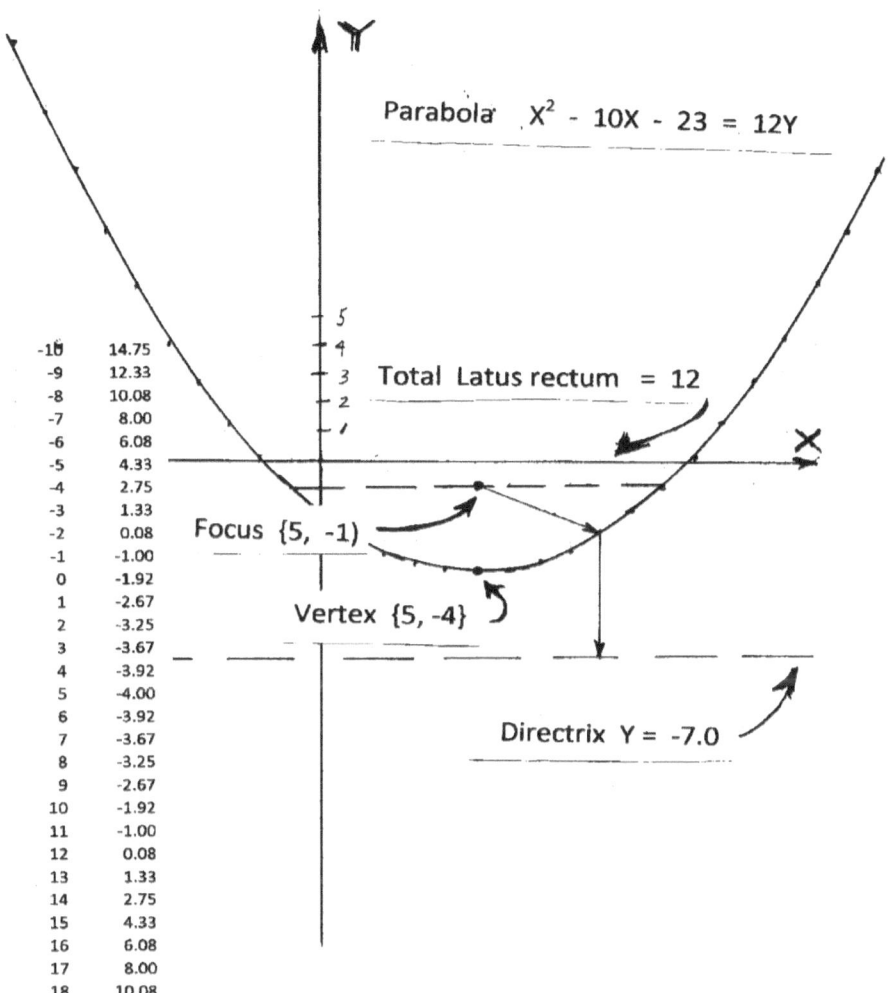

Chapter 83.2 Derivatives for Parabola

Use parabola on page 203 as an example.

$12 Y = X^2 - 10X - 23$

We want to find the slope at any point on the parabola.

The slope at any point is the rise divided by the height or $\Delta Y / \Delta X$ where Δ is a very small distance from point X and Y.

Replace X by $X + \Delta X$ and Y by $Y = \Delta Y$.

$12 (Y = \Delta Y) = (X + \Delta X) - 10 (X + \Delta X) - 23$

Since $12Y = X^2 - 10X - 23$

We can write $12 \Delta Y = 2X + (\Delta X)^2 - 10 \Delta X$

Divide equation by ΔX $12 \Delta Y / \Delta X = 2X + \Delta X - 10$

Since $\Delta X \ll X$; we can write $12 \Delta Y / \Delta X = 2X - 10$

At the minimum point $\Delta Y / \Delta X = 0$ (slope = zero)

Then $X = 5$ which agrees with the location of the vertex of the parabola

Chapter 83.3 Factoring $1000C^3 - 27$

$(C - 0.3)$ is one factor and a solution to $Y = \mathbf{1000C^3 - 27}$

Using synthetic division

$$
\begin{array}{r}
C^2 + 0.3C + 0.09 \\
C - 0.3 \overline{\smash{)}\, C^3 + 0C^2 + 0C + 0.027} \\
\underline{C^3 - 0.3C^2 } \\
0.3C^2 - 0C \\
\underline{0.3C^2 - 0.09C } \\
0.09C - 0.027 \\
\underline{0.09C - 0.027}
\end{array}
$$

$(C-0.3)\,(C2 + 0.3C + 0.09)$ are the factors

Using the quadratic formula, the second factors become

$\left\{C + \dfrac{0.3 + 0.3\sqrt{-0.3}}{2}\right\} \left\{C + \dfrac{0.3\sqrt{-0.3} - 0.3}{2}\right\}$ $\{C - 0.3\}$

Factoring $64 - a^3$

Set $64 - a^3 = 0$; $a = 4$, then $a - 4$ is one factor

Use synthetic division to find two other factors.

A cubic polynomial has three roots and three factors.

$$
\begin{array}{r}
16 + 4a + a^2 \\
4-a \overline{\smash{\big)}\, 64 + 0a + 0a^2} \\
\underline{64 - 16a } \\
16a + 0a^2 \\
\underline{16a - 4a^2 } \\
4a^2 - a^3 \\
\underline{4a^2 - a^3}
\end{array}
$$

Roots are: $(4-a)(16 + 4a + a^2)$ Using the quadratic formula;

$(4 - a) \left(a\, \dfrac{-4 + 4\sqrt{-3}}{2} \right) \left(a\, \dfrac{-4 - 4\sqrt{-3}}{2} \right)$

Chapter 83.4 Adding integers in a sequence

1 = integer, N = final integer in sequence

N = 8 is arbitrary end of the sequence

i = indicies 1, 2, 3, 4, 5, 6, 7, 8.......N

Sum of integers = 1 + 2 + 3 + 4 + 5 + 6 + 7 + 8 = 36

$\sum i$ from i = 1 to i = N = (N + 1) N/2 = (8)(9)/2 = 36

Sum of integers^2 = $1^2 + 2^2 + 3^2 + 4^2 + 5^2 + 6^2 + 7^2 + 8^2$ = 204

$\sum i^2$ from i = 1 to i = N = (N)(N +1)(2N +1)/6 = (8)(9)(17)/6 = 204

Sum of even integers = 2 + 4 + 6 + 8 +10 + 12 +14 +16 = 72

$\sum 2i$ (even) from i = 1 to i = N = (N +1) N = (8) (9) = 72

Sum of odd integers = 1 + 3 + 5 + 7 + 9 +11 +13 + 15 = 64

$\sum i$ (odd) from i = 1 to i = N = $N^2 = 8^2$ = 64

Sum of integers^3 = $1^3 + 2^3 + 3^3 + 4^3 + 5^3 + 6^3 + 7^3 + 8^3$ = 1296

$\sum i^3$ from i = 1 to i = N = $[(N)(N+1)/2]^2 = [(8)(9)/2]^2$ = 1296

Chapter 84.0   References

Code Book by Simon Singh

The Man Who Loved Only Numbers by Paul Hoffman (Biography of mathematician Paul Erdos)

RSA Web Site and PGP Web Site, Cryptologic Publication 1-0

In Code (Biography of mathematician Sarah Flannery)

Asier Technologies Data Sheet

"Quantum Cryptography" by Bennet, Brassard, & Ekert, *Scientific American*, Oct. 92

The Code Breakers by David Kahn Revised Edition 1996

Quantum Computing with Molecules by Gershenfeld and Chuang June 1998

Scientific American

Wikipedia

How Math Explains the World by James. D. Stein

Student Calculator Math by Texas Instruments

Computational Mathematics by Shlomo Breuer and Gideon Zwas

Mathematics of Physics and Modern Engineering by Sokolnikoff and Redheffer

Private conversation with Naomi (Breuer) Greenbaum, daughter of Professor Shlomo Breuer of Blessed Memory.